Palgrave Series in Indian Ocean World Studies

Series Editor
Gwyn Campbell
Indian Ocean World Centre
McGill University
Montreal, QC, Canada

This is the first scholarly series devoted to the study of the Indian Ocean world from early times to the present day. Encouraging interdisciplinarity, it incorporates and contributes to key debates in a number of areas including history, environmental studies, anthropology, sociology, political science, geography, economics, law, and labor and gender studies. Because it breaks from the restrictions imposed by country/regional studies and Eurocentric periodization, the series provides new frameworks through which to interpret past events, and new insights for present-day policymakers in key areas from labor relations and migration to diplomacy and trade.

More information about this series at
http://www.palgrave.com/gp/series/14661

Gwyn Campbell • Eva-Maria Knoll
Editors

Disease Dispersion and Impact in the Indian Ocean World

palgrave
macmillan

Editors
Gwyn Campbell
Indian Ocean World Centre
McGill University
Montreal, QC, Canada

Eva-Maria Knoll
Institute for Social Anthropology
Austrian Academy of Sciences
Vienna, Austria

Palgrave Series in Indian Ocean World Studies
ISBN 978-3-030-36263-8 ISBN 978-3-030-36264-5 (eBook)
https://doi.org/10.1007/978-3-030-36264-5

This research was supported by the Social Sciences and Humanities Research Council of Canada.

Cover illustration: GraphicaArtis/Getty Images

This Palgrave Macmillan imprint is published by the registered company Springer Nature Switzerland AG.
The registered company address is: Gewerbestrasse 11, 6330 Cham, Switzerland

Contents

Notes on Contributors

Edward A. Alpers is Research Professor of History at the University of California, Los Angeles. He has also taught at the Universities of Dar es Salaam, Tanzania (1966–1968) and the Somali National University, Lafoole (1980). In 1994 he served as President of the African Studies Association (USA). Alpers has published widely on the history of East Africa and the Indian Ocean. His major books include *Ivory and Slaves in East Central Africa* (1975), *Africa and the West: A Documentary History from the Slave Trade to Independence*, with William H. Worger and Nancy Clark (2001, 2nd ed. 2010), *East Africa and the Indian Ocean* (2009) and *The Indian Ocean in World History* (2014). He has co-edited *Walter Rodney: Revolutionary and Scholar* (1982), *History, Memory and Identity* (2001), *Sidis and Scholars: Essays on African Indians* (2004), *Slavery and Resistance in Africa and Asia* (2005), *Slave Routes and Oral Tradition in Southeastern Africa* (2005), *Resisting Bondage in Indian Ocean Africa and Asia* (2007), *Cross-Currents and Community Networks: The History of the Indian Ocean World* (2007), *Changing Horizons of African History* (2017), *Connectivity in Motion: Island Hubs in the Indian Ocean World* (Palgrave Macmillan, 2018) and *Transregional Trade and Traders: Situating Gujarat in the Indian Ocean from Early Times to 1900* (2019). He is senior editor for the *Oxford Research Encyclopedia of Asian History and Associate*, editor of the *Oxford Encyclopedia of Slavery, the Slave Trade, and the Diaspora in African History.*

Gwyn Campbell is founding Director of the Indian Ocean World Centre at McGill University, general editor of the *Palgrave Series in Indian Ocean World Studies* and editor-in-chief of the *Journal of Indian Ocean World Studies* (*JIOWS*). Born in Madagascar, and raised in Wales, he holds degrees in economic history from the Universities of Birmingham and Wales and has taught in India (Voluntary Service Overseas) as well as at universities in Madagascar, Britain, South Africa, Belgium and France. He served as an academic consultant for the South African Government in a series of inter-governmental meetings that led to the formation of an Indian Ocean regional association in 1997. He currently holds a Humboldt Award (2017–2019) for his research and teaching in Indian Ocean world studies and is director of a major international research project entitled "Appraising Risk, Past and Present: Interrogating Historical Data to Enhance Understanding of Environmental Crises in the Indian Ocean World." His publications include *Africa and the Indian Ocean World from early times to 1900* (Cambridge University Press, 2019), *David Griffiths and the Missionary "History of Madagascar"* (2012), *An Economic History of Imperial Madagascar, 1750–1895* (Cambridge University Press, 2005) and, as editor, *Bondage and the Environment in the Indian Ocean World* (Palgrave Macmillan, 2018) and *Africa and the Early Indian Ocean World Trade to circa 1300* (Palgrave Macmillan, 2016).

Manikarnika Dutta completed her MSc in the History of Science, Technology and Medicine on a Wellcome Trust Master's Studentship from the University of Oxford. She is currently pursuing a Wellcome Trust-funded DPhil project from the same institution. Her research examines the health and sanitary regulation of European seamen in colonial Indian port cities, integrating the history of health, imperial governance, maritime exchange and public policy in the British Empire. She was awarded the Taniguchi Medal (2018) by the Asian Society for the History of Medicine for best graduate essay submission.

Monica H. Green is a historian of medicine and health. She has worked throughout her career in the field of mediaeval European medical history, focusing primarily on questions of social and intellectual history. Recent developments in the fields of bioarchaeology and genetics have enabled new approaches for the reconstruction of the history of the world's leading infectious diseases, including the mediaeval scourges, plague and leprosy. These have induced Green to expand her teaching and research into Global History. Her recent publications include *Pandemic Disease in the Medieval World: Rethinking the Black Death* (editor, 2014/2015), "Putting Africa on the Black Death Map: Narratives from Genetics and History" (*Afriques* 9/2018) and "Richard de Fournival and the Reconfiguration of Learned Medicine in the Mid-13th Century," in: *Richard de Fournival et les sciences au XIIIe siècle*, edited by Joëlle Ducos and Christopher Lucken (2018). Together with Nükhet Varlık and Joris Roosen, she is currently developing a multidisciplinary *Black Death Digital Archive*.

Peter Hynd is a PhD candidate in the Department of History at McGill University, where he has worked as project manager of McGill's Indian Ocean World Centre. He holds a Master's Degree in History from McMaster University in Hamilton, Ontario, Canada, and an undergraduate degree in history from the University of Toronto. His research focuses on the regulation and taxation of alcohol in British India during the nineteenth century. He has also published on the history of cannabis in India, including a chapter in the forthcoming *Cannabis: Global Histories* (2021).

Karine Aasgaard Jansen is a senior researcher at the Department of Culture and Media Studies at Umeå University, Sweden. Her primary research interests are within the field of medical anthropology, particularly how public health discourses and interventions feed into local illness experiences and conceptualizations of vector-borne epidemics. She is currently leading the research project "Contagion and Culture: The 2005 to 2007 chikungunya epidemic in the Western Indian Ocean" funded by the Swedish Research Council. The study focuses on human–environment interaction and its effect on the diffusion and understanding of chikungunya across the islands of Réunion and Mauritius. Her most recent publications are from her postdoc project on the swine flu pandemic and subsequent mass vaccination in Scandinavia and include "What to expect when you're expecting a pandemic: Public health and lay perceptions of the 2009–2010 swine flu outbreak and mass vaccination in Norway", *Ethnologia Scandinavica* 48, 2018, and "To be vaccinated, or not to be vaccinated, is that the (only) question? Norwegian perceptions of vaccination and the swine flu pandemic in 2009 and 2010", *Tidsskrift for Kulturforskning*, 17/1, 2018. She has also contributed to the anthology *Histories of Medicine and Healing in the Indian Ocean World*, edited by Facil Tesfaye and Anna Winterbottom (Palgrave Macmillan, 2016), with a chapter entitled "Tropical disease and the making of France in Réunion."

Lori Jones is a postdoctoral fellow in the Department of History at Carleton University (funded by the Social Sciences and Humanities Research Council of Canada and previously by Associated Medical Services) and a sessional professor at the University of Ottawa. She received her PhD in History from the University of Ottawa in 2017, where her dissertation research focused on the evolution of ideas about the history and geography of the plague as seen through medical treatises. Her postdoctoral research addresses the interplay between manuscript and print versions of medical texts in the early modern era. Her publications include "Unrecorded Versions of John of Burgundy's Plague Tract and Identifying 'Lost' Copies of the Same," *Notes and Queries* 65/1, 2018; "The Diseased Landscape: Medieval and Early Modern Plague-Scapes," *Landscapes* 17/2, 2016 and (with Richard Nevell) "Plagued by Doubt and Viral Misinformation: The Need for Evidence-based Use of Historical Disease Images," *Lancet Infectious Diseases* 16/10, 2016. In addition to several in-progress articles and book chapters, she is currently wrapping up a monograph version of her dissertation and is editing two upcoming volumes: *Disease and the Environment in the Medieval and Early Modern Worlds* and (with Nükhet Varlık) *Death and Disease in the Medieval and Early Modern World: Perspectives from Across the Mediterranean and Beyond.*

Eva-Maria Knoll is a researcher at the Institute for Social Anthropology, Austrian Academy of Sciences. Her research interests focus on medical anthropology at the intersection with life sciences, health-related mobility and tourism. Currently she investigates the biosocial impact of an inherited blood disorder in the Republic of Maldives and she works on the challenge of hemoglobinopathies as rare diseases in Austria. She has co-edited *Camels in Asia and North-Africa* (2012). Her recent publications include "The Maldives as an Indian Ocean Crossroads", in: *Oxford Research Encyclopedia of Asian History* (2018), "Considering the Island Capital Male' as a Hub for Health-related Mobilities," in: *Connectivity in Motion: Island Hubs in the Indian Ocean World,* edited by Burkhard Schnepel and Edward A. Alpers (Palgrave Macmillan, 2018) and "Archipelagic Genes: Medical Travel as a Creative Response to Limitations and Remoteness in the Maldives", *Asia Pacific Viewpoint* 58/2, 2017.

Eric A. Strahorn received his PhD from the University of Iowa in 1997, where his dissertation examined the place of disease, wildlife and agriculture in the human landscape of the tarai region of north India. He is an associate professor at Florida Gulf Coast University and has been a research fellow of the National Asia Research Program of the National Bureau of Asian Research and Woodrow Wilson International Center for Scholars, a Florence Tan Moeson Fellow at the Library of Congress and a Fulbright Fellow. His first book *An Environmental History of Postcolonial Northern India: The Himalayan Tarai in Uttar Pradesh and Uttaranchal* was published in 2009 (Peter Lang) and "The Indus River Basin in the 21st Century" in Robert M. Hathaway and Michael Wills (eds.) *New Security Challenges in Asia* was published in 2013.

James Francis Warren is Emeritus Professor of Southeast Asian Modern History at Murdoch University, Perth, Western Australia. He has held positions at ANU, Yale University and as a professorial research fellow at the Centre for Southeast Asian Studies, Kyoto University, and the Asia Research Institute, National University of Singapore. He has been awarded grants by the Social Science Research Council and the Australia Research Council and is a fellow of The Australian Academy of the Humanities. He is currently the director of a major Australia Research Council Linkage Project, "Hazards, Tipping Points, Adaptation and Collapse in the Indo-Pacific World." Professor Warren's major publications include *The Sulu Zone, 1768–1898* (1981), *Rickshaw Coolie: A People's History of Singapore, 1880–1940* (1986), *At the Edge of Southeast Asian History* (1987), *Ah Ku and Karayuki-San: Prostitution and Singapore Society, 1870–1940* (1993), *The Sulu Zone, the World Capitalist Economy and the Historical Imagination* (1998), *Iranun and Balangingi: Globalization, Maritime Raiding and the Birth of Ethnicity* (2002) and *Pirates, Prostitutes and Pullers: Explorations in the Ethno- and Social History of Southeast Asia* (2008). In 2003, he was awarded the Centenary Medal of Australia for service to Australian society and the Humanities in the study of Ethnohistory, and in 2103, the Association for Asian Studies, Grant Goodman Prize in Historical Studies.

Anna Winterbottom is a research associate in an SSHRC-funded grant at McGill focusing on nineteenth-century natural history in India. She previously held postdoctoral positions at McGill and Sussex University. Her interests are in the early modern Indian Ocean region and the European colonial presence there, with a focus on the history of medicine, science and environment. Her current book project focuses on multi-directional exchanges of *materia medica* and healing objects, ideas and practices around the Indian Ocean region between c. 1500 and 1800

CE. She is the author of *Hybrid Knowledge in the Early East India Company World* (Palgrave Macmillan, 2016) and the co-editor of *Histories of Medicine and Healing in the Indian Ocean World* (Palgrave Macmillan, 2015) and *The East India Company and the Natural World* (Palgrave Macmillan, 2014). In 2017, she won the J. Worth Estes Prize of the American Society for the History of Medicine for her journal article, 'Of the China Root: a Case Study of the Early Modern Circulation of Materia Medica', published in *Social History of Medicine* 28/1, 2015.

List of Figures

List of Charts

List of Maps

List of Pictures

List of Tables

CHAPTER 1

Introduction

Eva-Maria Knoll and Gwyn Campbell

Environment and Movement

Throughout history in the Indian Ocean World (IOW) diseases have, under certain distinctive geographical and climatic conditions, emerged and spread, generating a number of impacts on varying spatial scales. The IOW, a macro-region lying between latitudes 45°S and 45°N running from Eastern Africa through the Middle East, South and Southeast Asia to East Asia, encompasses tropical, sub-tropical, and temperate zones, major oceans, gulfs and rivers, islands, lakes and deserts, and the world's highest mountain range (Map 1.1). It thus experiences major differences in temperature and rainfall, which are further affected by other environmental factors—the most significant of which is the monsoon system of winds and currents that governs the IOW littorals and seas north of about 12°S of the equator. In the northern hemisphere summer, the southwest monsoon dominates, bringing heavy rainfall to the Asian littoral, while in winter the

E.-M. Knoll (✉)
Institute for Social Anthropology, Austrian Academy of Sciences, Vienna, Austria
e-mail: Eva-Maria.Knoll@oeaw.ac.at

G. Campbell (✉)
McGill University, Montreal, QC, Canada
e-mail: gwyn.campbell@mcgill.ca

G. Campbell, E.-M. Knoll (eds.), *Disease Dispersion and Impact in the Indian Ocean World*, Palgrave Series in Indian Ocean World Studies, https://doi.org/10.1007/978-3-030-36264-5_1

system switches direction, creating the northeast monsoon. Most historians have assumed the monsoon system to have been stable, but it could unpredictably fail, triggering drought, crop failure, famine, and disease. A range of other, often associated, environmental factors, such as the El Niño–Southern Oscillation (ENSO), Indian Ocean Dipole (IOD), Intertropical Convergence Zone (ITCZ), volcanism, and cyclones could also significantly impact temperature and rainfall and thus disease. For example, in the aftermath of heavy rain, stagnant pools of water could form, providing breeding grounds for mosquitos and other causal agents of diseases such as malaria, filariasis, dengue, and chikungunya (cf. Meunier 2014). Again, heavy monsoons, cyclones, seismic activities, tsunamis, and storm surges could lead to flooding that might in turn create favourable conditions for pathogenic microorganisms and thus for the spread of water-borne and contagious diseases such as cholera, dysentery, and polio or *Escherichia coli* infections. Furthermore, weather extremes and natural disasters were often followed by famines, conflict, and migration, all of which increase health hazards.

In addition to these environmental specificities—this "deep structure" of the IOW (cf. Pearson 2003)—the macro-region witnessed the rise of the first "global" economy from around 300 BCE. The IOW global economy, linking Eastern Africa and the Middle East to China and all places in between through the creation of a sophisticated network of overland, riverine, and maritime communication, was characterized by an intensifying exchange of plants, animals, and (both voluntary and involuntary) humans—creating the quintessential conditions for disease diffusion. This process, which started with early hominid migration out of Africa, triggered the development of regionally specific immunological responses. With the advent of long-distance trans-IOW seafaring, the entanglement of humans and pathogens gained a novel epidemiological momentum affecting both littoral and hinterland communities (cf. Campbell 2019; Schnepel and Alpers 2017; Pearson 2015; Alpers 2014; Sheriff 2010). The interconnected character of the IOW global economy, and increasing concentration of human and animal populations close to water resources, transformed the IOW into one interconnected disease zone (Issa 2006; Arnold 1991). It formed, for example, a centre of dispersion of a number of diseases such as the plague, smallpox, malaria, and tuberculosis.

However, disease outbreaks and dispersion did not occur in a historically linear fashion. The IOW global economy underwent major cycles of expansion and contraction. The main eras of economic expansion were from approximately 300 BCE to 300 CE, between the ninth and thirteenth centuries, and again from the mid-nineteenth century—the intervening periods being marked by general stagnation. Times of overall economic prosperity, characterized by enhanced agricultural productivity and demographic growth, were not immune from outbreaks of disease. However, it is notable that some of the most notable and devastating episodes of disease, such as the first and second plague (541 and 1347 CE) and cholera (e.g. 1817 and 1826 CE) pandemics, not only originated in the IOW but erupted during periods of major economic uncertainty (Campbell 2019).

Within this context, there is considerable debate about the European impact on the IOW. For some scholars, such as Arnold, the advent of the European presence from 1500 marked a major epidemiological watershed for the IOW (Arnold 1991). However, Campbell argues that, in contrast to the New World to which Europeans carried Old World diseases that had a catastrophic impact on indigenous populations, the reverse was generally true in the IOW where Europeans proved highly vulnerable to tropical diseases. This was, for example, the case with malaria to which many African populations had, through genetic adaption (sickle cell), acquired resistance. Thus the Portuguese in Mozambique suffered such high mortality from malaria that they often lacked sufficient soldiers to maintain a garrison. For largely the same reason, European attempts to found settlements in Madagascar failed. Only with the widespread adoption of quinine from the late nineteenth century could European soldiers and colonists settle malarial regions of the IOW (Campbell 2019 and contribution to this volume).

The nineteenth century marked a major turning point in the disease history of the IOW for a number of reasons, many of which were related to the rise of a truly international economy that, by the eve of the First World War, had drawn all bar the most peripheral societies into the orbit of modern capitalism. First, with the exception of railways in India, and of late nineteenth-century investment in mining in South Africa, few areas of the IOW benefitted from the enormous outflow of capital to extra-European regions from financial centres, notably London and Paris, based in Western Europe. Consequently, growing demand for tropical and semi-tropical products from a rapidly industrializing West resulted in a commercial boom in the IOW that was largely labour intensive. This was the case with both European and indigenous enterprise in the macro-region. Manpower was required to clear land and cultivate cash crops such as

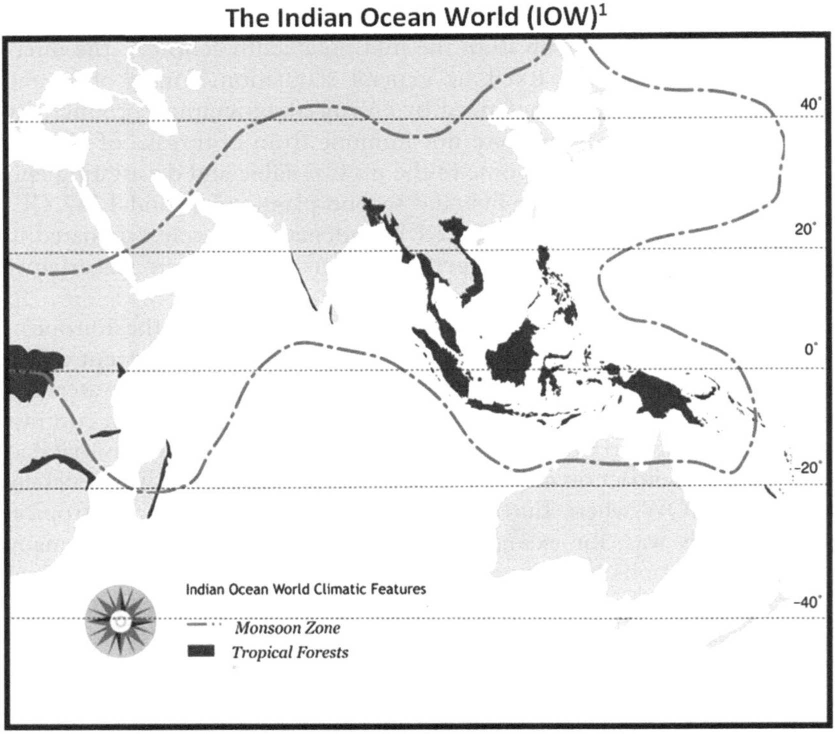

Map 1.1 The Indian Ocean World (IOW). © IOWC

cloves, sugar cane, coffee, tea, and cocoa; collect forest products such as gum and tropical hardwoods; hunt and extract prized animal products such as ivory, rhino horn, skins, pearls, and whale oil; transport such produce to ports, and carry imported articles into the interior; load and offload commodities at ports; and provide the crews of commercial vessels. However, in the 1800s the IOW failed to experience the same rates of demographic expansion as the West, and, as much labour was already tied up in indigenous forms of bondage, there existed a very limited wage labour force from which to hire workers. As a result, European and indigenous IOW authorities, traders, and entrepreneurs, resorted largely to the use of bonded labour. This was reflected in the continued use of slaves and the increased use of penal and especially of indentured labour. There developed a large-scale system of bonded labour movements, both intra-IOW and from the IOW to labour markets in other regions of the world, such as the Caribbean, characterized by labour shortages. These large-scale long-distance movements, which continued well into the twentieth century, brought immunologically naïve people into contact with previously unknown diseases and environmental conditions. Such mobile or displaced people, in turn, carried diseases, predispositions for certain conditions, and specific immunological responses to new locations (Campbell 2018, 2014; Sheriff and Ho 2014).

Additionally, the nineteenth and twentieth centuries were marked by major imperialist ventures, both European and indigenous, in which there occurred massive movements of troops and camp followers. Military expeditions often led to significant displacements of populations, both combatant and non-combatant. At the same time, the presence of major concentrations of male soldiers and workers resulted in a rise in the transmission of sexually transmitted diseases (STDs) and of alcoholism. First, such concentrations of males (few European females accompanied white soldiers to the IOW, and indentured labourers were predominantly male—overwhelmingly in the case of the Chinese) established a huge demand for sexual services. This in turn resulted in an often officially encouraged system of recruitment and migration of young women from mostly impoverished backgrounds, who were often forced into the provision of such services. Inevitably, these systems resulted in an explosion in the transmission and diffusion of sexually transmitted diseases. Massive concentrations of male soldiers and workers also led to an unprecedented rise in, and growing public concern about, alcohol consumption. This theme is explored in this volume by Peter Hynd and Manikarnika Dutta who reveal that, in nineteenth-century India, alcohol was a highly lucrative trading commodity that some officials considered to be a panacea for certain deadly diseases such as cholera and the bubonic plague. Official tolerance of the production and sale of alcohol encouraged European seamen and soldiers to consume it, often in adulterated form, on unprecedented levels. However, missionaries in the field, and members of a burgeoning temperance movement in Western Europe and North America, argued with increasing force that alcohol was hazardous to health and morality.

The nineteenth and twentieth centuries were also characterized by an ever accelerating series of technological advances that further propelled disease dispersion. This was particularly the case with transport innovations, notably the expansion in the use of railways from the mid-nineteenth century, steamships from the 1870s, civil aviation from the 1930s, container and liner shipping from the 1950s, and long-haul flight mass tourism and travel in the modern era (cf. Alpers 2014; Mitchell 2016).

Disease in IOW History

Disease has played a crucial role in shaping the size, movement, and settlement patterns of human populations as well as their peaceful or conflict-driven encounters. Moreover, thinking about and acting upon disease is central to all world views and thus provides a window into how societies perceive events and their progression. Hence, disease serves as an indispensable cornerstone in the reconstruction of human history. Disease is experienced, socially understood, and governed, in the broader understanding of Foucault (2011), as comprising governmental interventions as well as techniques of self-control. This allows for a wide range of investigative approaches that include the addressing of individual experiences of illness, the social organization of care, and institutional interventions.

This volume views the study of disease as essential to an understanding of the key historical developments underpinning the foundation of contemporary IOW societies. In this exercise, it is necessary first to define what we mean by "disease" and second to examine the exchange dynamics of pathogens and healing techniques across both the terrestrial and maritime zones of the IOW. We here follow historian Andrew Cunningham's identification of three fundamental dimensions of disease as "(1) an *experience*—an experience of debilitation, pain, suffering, together with (2) the spontaneous *appearance of non-customary phenomena* with respect to the body, such as spots, vomiting, sweating, aches, and (3) … *outcomes* of recovery, death or disability" (Cunningham 2002, 13). Furthermore, humans "seem to insist on seeking reasons or *causes* for disease: for its incidence, its origin, its course, its outcome" (ibid.).

The origins and routes of disease in the IOW were both terrestrial and maritime. The fourteenth-century pandemic of bubonic plague, commonly called the Black Death, offers a prime example of the terrestrial origins and diffusion of disease. There exist three endemic foci of rodent populations that carry *Yersinia pestis*, the organism that causes bubonic plague: the Eurasian steppe (between Manchuria and the Ukraine), Himalayan foothills, and Great Lakes region of East Africa. The second plague pandemic started probably in China in the 1330s. It erupted because of two occurrences: an epizootic of the plague amongst the rodent population and sustained contact between the affected rodents and humans. These were probably coterminous and related to major environmental events that upset the habitat and challenged the immune system of the rodents. First, from about 1300, China entered a prolonged period of economic and political turbulence. Second, East and Southeast Asia experienced major climatic and environmental disturbances. In China, the early 1330s were characterized by excessive rain and flooding, while in 1334, Mount Kelud in Eastern Java erupted with a volcanic flux magnitude comparable to the 1815 Tambora eruption ($f \approx 26.3$ kg km^{-2}). Such eruptions resulted in reduced temperatures for two to three years, and harvest shortfalls, often accompanied by famine and disease. During these early fourteenth-century events, an epizootic of plague erupted amongst a population of rodents, forcing them to flee their natural habitat and seek refuge and food in communities of humans, to which the pathogen-carrying fleas transferred from the dying rodents. The disease affected visiting traders, subsequently travelling with some of them along the overland Silk Road to the Near East and Europe: this route ran from China's Mongolian border across Central Asia to Neyshabur (Nishapur), in North-Eastern Iran, from where major trade routes ran to Northern India, Mesopotamia, and on to the Persian Gulf or the Mediterranean, and north via the Caspian Sea to Russia. At the pace of 30 km a day, it would have taken such traders and the plague just over four months to travel from China's Mongolian frontier to Neyshabur. Approximately halfway lies Lake Issyk-Kul, near which are located two mediaeval cemeteries of Nestorian Christian traders. These reveal that at least 106 of the 650 people buried there between 1186 and 1349 died in 1338–1339—and that "pestilence" was marked as the cause of death of at least 10 of them. It is estimated that in Europe, the Black Death killed one third or more of the population (Campbell 2019; Reid 2018; Bos et al. 2016; Sussman 2011; Cohn and Weaver 2006).

Whereas human diseases emanated overwhelmingly from land-based sources, the development of transoceanic voyaging ensured that ships, their cargoes, crews, and passengers also constituted highly significant factors in the long-distance transmission of disease. Most scholarly work on the bio-cultural history of intentional and unintended transoceanic exchange of diseases has focused on the Atlantic in the era of European expansion (e.g. Crosby 1972). Only following David Arnold's seminal 1991 article emphasizing the epidemiological distinctiveness of the Indian Ocean have the maritime spaces of the IOW entered the academic discourse on disease zones of exchange and transformation (Arnold 1991). The early domestication of animals, development of densely populated trading hubs, and rise of trans-IOW oceanic sail allowed for germ exchange across the maritime spaces of the macro-region from about 300 BCE—well before the European incursion into the IOW from around 1500. Arnold, whose focus was specifically on the Indian Ocean, stressed the development of "epidemic highways" (ibid., 4).

Not least of these were criss-crossing pilgrim routes, notably those of the Muslim hajj to Mecca, which constitutes "one of the greatest, and longest lasting, maritime passenger traffics in the world" (Pearson 2015, 9). Due to the IOW's deep pre-colonial historical interconnectedness, the European impact was not as abrupt and catastrophic as in the New World (Arnold 1991, 5). Nevertheless, as Arnold points out, "the emergence of India as the lynch pin of British power and trade in the East was of great epidemiological significance for the rest of the region and indeed the wider world beyond" (ibid., 7). In the contemporary era, the IOW has witnessed the eruption in epidemic and pandemic forms of a number of diseases such as dengue, chikungunya, SARS, Zika, and subtypes of influenza (Weaver and Lecuit 2015; Zeller 1990; Alpers, Jansen this volume).

The Role of Islands

Of notable significance in disease formation and distribution processes in the IOW were islands (cf. Falola et al. 2019; Pearson 2003; Alpers 2000, 2014)—the focus of five of the chapters in this volume (i.e. Alpers, Campbell, Jansen, Knoll, Warren). Taking advantage of the "laboratory" quality of island settings (cf. Cliff et al. 2000), these contributions reveal in condensed form processes of disease formation, dispersion, and management. As tracts of land surrounded by water, islands are separated, even isolated, as well as bridged and connected by water. Islands thus were both convenient outposts on the peripheries of countries, societies, and empires, and indispensable nautical nodal points in complex maritime networks. As crucial stopovers in trade, stepping stones in migratory movements, and gates to continental hinterland resources and power, islands were also, inadvertently, pivotal centres of endemic disease formation, of virulent epidemic invasion, and of the exchange and expansion of diseases.

Indeed, islands were overexposed to specific health hazards. Smaller islands, often densely populated, were vulnerable to resource scarcity, seasonal storms, drought, and flooding—which often created unique disease environments that challenged the human immune system. Plantation projects on larger islands such as Java and the Philippines could have lasting impacts on human health and wellbeing. Some islands, such as the Maldives and Madagascar, even gave their names to supposedly distinctive fevers. In addition to ecological factors, islands could be precariously dependent on sea traffic and the changing fortunes of the IOW global economy. Some islanders, notably those in pivotally strategic locations along the IOW maritime networks, such as coastal Ceylon, Sumatra, and Java, as well as on smaller islands such as Anjouan and Mahé, were in early and regular contact with voyaging seafarers, traders, and travellers. At the centre of inter-regional and cross-IOW exchange, they were shaped by cosmopolitanism and sexual relations across racial and ethnic boundaries. Islands were distinctive disease dispersion hubs. Genetic founder effects weighed heavily on small populations, and epidemics travelled fast through close-knit island communities. Moreover, the enduring turbulent history of many IOW islands led to the juxtaposition of multi-ethnic and multi-religious populations with corresponding social tensions and conflict dynamics, as is conspicuous, for example, in the case of Sri Lanka.

Their geographical location, and historical circumstances, made some islands epidemiological avenues to hinterlands and staging posts along the epidemic highways in the IOW disease zone. Domination of islands and littorals was thus about controlling not only trading routes but also the pathways and speed of epidemics. Discrete island spaces that existed in abundance in many parts of the Indian Ocean, and were generally manageable due to their small size, served also as liminal spaces, usable, for example, as quarantine islands or leper colonies.

Early Disease Control and Treatment

One of the major human reactions to disease in the IOW was the development and exchange of therapeutic efforts of disease control and treatment (Issa 2006), and theoretical considerations of disease cause and transmission. The IOW is the birthplace of some of the world's oldest medical systems, and a space of intensive exchange of healing techniques, materials, and knowledge. This included the exchange and interlocking dynamics of therapeutics, *Materia medica*, and skilled practitioners of the Ayurveda, Chinese, Tibetan, Yunani (Arab-Persian) medical systems, as well as of Prophetic and Islamic medicine, and non-textual healing rituals and practices (Winterbottom and Tesfaye 2016). Moreover, the IOW was a laboratory for emerging scientific fields such as virology, malariology, parasitology, and tropical medicine. Major biomedical breakthroughs occurred in the macro-region, including the discovery in India in 1897–1898, by future Nobel laureate Ronald Ross, of malaria transmission by the *Anopheles* mosquito. Quinine, the most widely used prophylaxis for malaria, appears in the IOW to have been first used in Madagascar (Campbell, this volume)—after Jesuits gained knowledge about the traditional medical use of the ground bark of cinchona trees by the Quechua, a people indigenous to Peru, Bolivia, and Ecuador, and transferred this knowledge to Europe in the later sixteenth century. The most recent form of health-related mobility in the IOW is the development of a thriving medical tourism industry with major centres in Singapore and Bangkok, and a correlating increase in medical travel by international patients (e.g. Knoll 2017).

Three decades after David Arnold's seminal article, and three years after the subject was further enhanced by a collection edited by Anna Winterbottom and Facil Tesfaye (2016), this volume sets out to shed further light on the dispersal and impact of disease in the IOW. Disease knows no political or man-imposed frontiers, and some pandemics, such as cholera in the nineteenth century, crossed the entire macro-region which, as noted above, runs from Eastern Africa in the West to China in the East. Indeed, cholera, like plague and some other diseases emanating from the IOW, spread beyond the IOW proper. Time wise, this volume covers the longue durée, from the arrival of early hominids in the IOW up to the twenty-first century. The dynamics of various diseases are here reflected against the backdrop of major transformations in the mobility of humans within the IOW and their interaction with disease. IOW mobility covers travel on foot, mules, camels, and horses and by everything from animal-drawn carts, railways, cars, buses, and sail and steam ships to aeroplanes. Moreover, medical history has, over the centuries, progressed from the study of miasma-based disease and, by the latter part of the nineteenth century, germ-based disease, to a post-Second World War focus on molecular and gene-based disease. The available sources for a reconstruction of IOW medical history have thus expanded, from purely symptomatic patient descriptions, to predominately anatomic definitions of disease (following the initial use of dissections during the Renaissance), to what Michel Foucault has called "the clinical gaze" in the mid-nineteenth century—that is, to ever more detailed test results focusing on disease as a composition of symptoms, rather than a reliance on the subjective descriptions of patients (Foucault 1973), and since the 1990s to DNA analysis and to ancient DNA (aDNA) sequencing.

Origins, Routes, and Impact of Disease

Debates over the origin, dispersion, and impact of disease form a central focus of this volume. In their chapter, Monica H. Green and Lori Jones reconstruct the pre-modern trajectories of malaria, tuberculosis, leprosy, smallpox, and plague in the IOW, although they caution that more ancient DNA needs to be retrieved in order to fully incorporate the macro-region into molecular narratives of global medical history.

Malaria is caused by the protozoan parasite *Plasmodium*, four species of which are responsible for the disease in humans: *P. vivax*, *P. falciparum*, *P. malariae*, and *P. ovale*. In this study, Green and Jones focus on the two most dangerous species, *P. vivax* and *P. falciparum*. The older vivax strain is associated with an Asian and the falciparum with an African centre of distribution. *P. vivax* may have passed, in modified form, from primates to humans. It possibly became most prevalent when the hominid migrants out of Africa became settled rice farming communities. *P. falciparum*, by contrast, appears to have originated as a human disease in Africa 10,000 years ago and to have diffused geographically chiefly through the slave trade: it spread throughout the Roman Empire and travelled via the trans-Atlantic slave trade to the Americas where it found particularly fertile ground on sugar and cotton plantations.

Tuberculosis seems to have an East African origin, although more virulent TB strains were introduced to the IOW with European colonialism. Smallpox was probably another quintessential IOW disease, originating from human-camel-rodent interactions in the Horn of Africa in the second millennium BCE. Those who survived initial attacks developed lifelong immunity to it, so smallpox could only thrive by moving as a "childhood disease" between communities. IOW trade networks provided it with the ideal means of dispersion. Again the slave trade may have played a pivotal role. Prolonged physical intimacy, such as that which existed in concentrated groups of slaves, seems to have been key to the transmission of the slowly replicating leprosy organism. Current academic thinking is that the plague had a Northern Eurasian origin, although there is an intriguing possibility that all three plague pandemics may have originated within the IOW: the first "Justinian" plague pandemic in East Central Africa (Campbell 2019) from where it spread via maritime routes to Ethiopia and Egypt, the second in Central Asia from where it spread via the Silk Road, and the third in Yunnan, China, from where it initially diffused slowly via overland routes and then became global through maritime routes.

In her chapter in this collection, Anna Winterbottom demonstrates commonalities and connectivity across the IOW in how people understood the "Frankish disease" and treatments for it. In Sub-Saharan Africa and Southeast Asia, references to the Frankish disease, considered to have been the same as the "Great Box" disease, or syphilis, in Europe, were rare or absent. The presence of forms of the malady caused by other strains of the bacterium *Treponema pallidum*, such as the non-venereal yaws, seems to have prevented these areas of the IOW from experiencing major epidemics of syphilis. Elsewhere in the IOW, the Frankish disease was generally believed to have been caused by close contact with foreigners. Tracing symptoms and epidemiological blaming across the IOW, Winterbottom demonstrates that in the IOW derivatives of the term *afrang* (from ancient Persian for "Frank"—i.e. the "Frankish disease") were used to refer to syphilis, the suspected transmitters of which were frequently Europeans and the prostitutes they frequented. During the colonial era, the earlier perception of the Frankish disease as originating with Europeans was reversed.

Some of the early treatments for the Frankish disease remained in limited local use, while others spread. A few, such as *Guaiacum* from Central America, *Smilax* or "China root", and mercury, became global remedies. Winterbottom concludes that medical cultures within and beyond the IOW shared some basic understanding of health as rooted in a balance of key bodily substances. The Frankish disease triggered changes in medical thinking in the IOW by associating concepts of contagion with specific groups of people.

Eric A. Strahorn, in his contribution to this volume, reassesses the debate about the early dispersion of leprosy in the IOW from the vantage point of fresh insights from palaeopathology. Bringing literary and archaeological evidence into dialogue with DNA sampling, Strahorn develops a pathology history of the progression of interpretations and transmission theories of "Hansen's disease", as leprosy is also called. The physical manifestations of lepromatous leprosy, the more severe of the two main types of Hansen's disease, which include extensive skin lesions, loss of extremities, and a collapse of the nose, may be identified in disease descriptions found in a number of ancient Chinese, Indian, Egyptian, and circum-Mediterranean texts. Such sources suggest the presence of leprosy in India by the second millennium BCE, and in ancient Egypt, the Roman

Mediterranean, and China by the third century BCE. Archaeological evidence supports this chronology. One influential modern hypothesis proposed that the armies of Alexander the Great carried leprosy from India to the Mediterranean. Critics, however, point to the particularly slow incubation period of *M. leprae* of three to ten years and suggest an alternative hypothesis—that leprosy travelled along age-old trade routes linking India and the Mediterranean. Strahorn considers that, since the diagnosis of leprosy in skeletal remains is complicated, and the majority of skeletons with leprosy-attributed bone damage have been found in Europe, DNA evidence has the potential to fill gaps in the literary and archaeological record. Nevertheless, the lack of aDNA samples from many parts of the IOW, notably India, still leaves many questions unanswered about the role of the IOW in the early dispersion history of leprosy.

In his chapter, James Francis Warren examines the health impacts of climate, weather, and colonialism in the Philippines from the sixteenth century. Consulting sources from the colonial and post-independence periods, Warren draws a picture of an archipelago afflicted by successive adversities. Natural disasters, such as typhoons and droughts, often triggered outbreaks of disease. Arid periods, especially during El Niño events, could lead to the drying up of wells and rivers with, as a result, limited and unsafe supplies of drinking water that often triggered the outbreak of disease. Again, floods, common during the rainy season, often precipitated health problems including coughs, fever and flu, water-borne diseases such as cholera, typhoid fever, amoebic dysentery, diarrhoea, and mosquito-borne diseases.

Cholera, which frequently erupted in the wake of typhoons and floods, was the most feared of water-borne diseases Seven major cholera outbreaks occurred in the Philippines during the nineteenth century, five of them linked to El Niño events. The Spanish colonial government relied heavily on the assistance of various Catholic religious orders and charitable donations for disaster relief and assistance. This proved insufficient, however, and what remained of the Spanish health system broke down under American rule from the late 1890s. In the post-colonial era, the Philippines have struggled with both ever accelerating population growth and the lack of adequate drinking water and sewage disposal systems. In poor rural and low-lying areas, as well as in urban agglomerations, cholera and typhoid remained major health threats. This was manifest, for instance, in the 1970s when the Philippines experienced a devastating series of typhoons and floods. Since then, rapid climate change, increased population density, and a growth in monocrop cash crops at the expense of food crops, have intensified the incidence and impacts of natural misfortunes and thus the vulnerability of the Filipino people to epidemics.

In his contribution, Gwyn Campbell examines the onset and spread of malaria in nineteenth-century highland Madagascar. "Madagascar Fever" early earned a notorious reputation amongst European visitors to the island and can be identified with reasonable certainty as malaria. Campbell shows that in Madagascar, the distribution and impact of malaria was determined by a mixture of climate, geography, and human activities. The disease probably arrived in the island with the first permanent settlers from East Africa in the eighth or ninth century CE. These pioneers comprised mixed or separate groups of Bantu speakers, who had sickle cell immunity to malaria, and Austronesians, who possessed no natural defences to malaria. This favoured the survival in Madagascar of Bantu speakers or those of mixed Bantu-speaking and Austronesian heritage who were sicklers. Later Austronesian arrivals (the proto-Merina) migrated to the central plateau chiefly to escape the largely endemic fevers of the lowlands. Attempts by Europeans, following their incursion into the IOW, to found settlements in the Malagasy lowlands largely failed due to malaria to which they proved highly vulnerable. The central highlands, at an altitude of 1300 to 1700 metres, were reputedly malaria free, but the evidence is that there, too, malaria became endemic in the course of the nineteenth-century, epidemic outbreaks occurring probably as early as the 1820s. This change in malaria epidemiology was connected to changing climatic conditions, notably warmer weather that facilitated the survival of mosquitos at higher altitudes, and to forced labour (*fanompoana*) policies that induced unprecedented flows of people between the malarial lowlands and the traditionally parasite-free highlands. This resulted in high mortality rates amongst highlanders. Also, people fled the land in order to avoid exploitative forced labour recruitment. In so doing, they abandoned labour-intensive rice fields and the water channels that fed them, clay-brick pits, and alluvial gold diggings, which created numerous reserves of stagnant water that provided ideal breeding ground for the malaria vector *Anopheles funestus* that spread from the eastern lowlands into the highlands.

In their contributions to this volume, Peter Hynd and Manikarnika Dutta examine the relationships between disease in colonial India, the production, regulation, consumption, and taxation strategies of alcohol, and colonial governance. In analysing the impact of disease in the context of alcohol consumption and social behaviour, these authors reveal two seemingly contradictory attitudes to alcohol which was, on the one hand, considered destructive of health, even a disease in its own right, and on the other was thought of as a remedy.

Focusing on Bombay Presidency, Peter Hynd explores the impact of the IOW disease environment, especially the major disease outbreaks of the late nineteenth century, on colonial governance. He examines official claims about the relationship between disease, alcohol consumption, and excise revenue, and evaluates these claims against mortality statistics. Scrutinizing recurring references to plague and other diseases in the official colonial excise records, Hynd demonstrates that, although disease did not have a serious impact on excise revenues in late nineteenth- and early twentieth-century India, excise department officials often invoked disease, especially in epidemic form, to justify striking increases or decreases in excise revenues. They argued, for example, that alcohol consumption decreased when people fled cities during disease outbreaks, such as the 1896–1897 plague epidemic that ravaged Bombay. The return of disease fugitives to the city after the crisis, and the widespread belief that liquor constituted a prophylactic against plague, probably led to increased alcohol consumption and thus a spike in excise revenues in 1897–1898. However, colonial excise agents were trapped between the imperative to enhance revenue and the increasingly critical voices of missionaries, temperance advocates, and nationalists, who denounced alcohol consumption. In colonial India, tax officials adopted a "maximum revenue from minimum consumption policy" to limit the drinking of alcohol, impose basic hygiene standards on distillers, and repress adulteration. Disease was a most welcome "explanation that worked" for the British Raj tax authorities to justify fluctuating excise revenues.

In her contribution, Manikarnika Dutta examines the impact of alcohol on the health and behaviour of sailors in nineteenth-century Calcutta. British authorities in India were worried about both the alcohol consumption of European sailors in Indian port cities and the low quality of local liquor. They were highly concerned about the drinking binges that sailors engaged in during shore leave and the "crimping system" whereby "crimps"—a particular term for a fraudster—enticed sailors to consume drugged liquor and subsequently tricked, mugged, and robbed their victims. Colonial officials started to investigate the quality of local liquor when reports proliferated of "treacherous" Indians tricking "innocent" European sailors into drinking "poisonous" liquor. Cheap, adulterated liquor was considered a health threat to white sailors and even envisaged to be a cause of cholera. Regulating the quality of liquor and its production and consumption facilities was therefore of paramount importance to British authorities. The temperance movement also campaigned against liquor consumption and related unruly activities, while the installation and administration of sailors' homes in port cities also helped to counter crimping and excessive drinking. Dutta investigates the extent to which the colonial state's measures to protect the health of European mariners were informed by imperial encounters in the fields of medical intervention, race relations, environmentalism, and legal order.

In his chapter, Edward A. Alpers extrapolates from a specific focus on the recent chikungunya epidemic in the western IOW to comment more generally about diffusion mechanisms of epidemic diseases. The RNA virus that causes *chikungunya* (a word used by the Makonde of Northern Mozambique and Southwest Tanzania meaning "that which bends up"—referring to the sufferer's severe joint pain) originated in two virus lineages in Africa and, in the late nineteenth century, developed a genetically distinct Asian virus genotype. The 2004 outbreak of chikungunya affected 75 percent of the population of Lamu Island, Kenya. It is thought that direct exchange between humans and mosquitos (i.e. independent of animal intermediaries), and intensified air travel, caused a rapid transmission of the disease, both to more distant island groups in the Western Indian Ocean such as Mombasa, the Comoros, the Mascarene Islands, and the Seychelles, and to India where in 2005–2006 over a million people were affected across 13 states. Mutations in the various chikungunya virus strains enabled additional mosquito species, such as the Asian tiger mosquito (*Aedes albopictus*), to become effective chikungunya vectors. This adaptation led to the re-emergence and rapid spread of chikungunya in the IOW some 50 years after the first authoritative identification of the disease in Tanganyika.

Alpers explores the academic debates surrounding the origins and recent expansion of chikungunya, including the virus-vector-environment dynamics of its aptly called Indian Ocean Lineage. The chikungunya pandemic demonstrates that this arthropod-borne virus can adapt and invade new hosts through mutation. With this insight, Alpers also reconsiders the malaria epidemics on the Mascarenes in the 1860s and poses three main questions for future research. First, could the *Anopheles* mosquito vector of the disease have survived a lengthy boat voyage from mainland Africa to these mid-ocean isolates? Second, did a critical mass of human malaria carriers then travel to these islands, enabling already existing populations of mosquitoes to become vectors and cause a malaria epidemic? And third, did a genetic mutation, similar to that which triggered the 2004–2007 chikungunya epidemic, transform existing mosquito populations into effective vectors for the *Plasmodium* parasite?

In her chapter, Karine Aasgaard Jansen examines the 2005–2007 chikungunya epidemic in Réunion, a French overseas department. As a social anthropologist, Jansen focuses on the social impact of this vector-borne disease in the context of local disease perceptions and the islanders' resistance to public health interventions. The Réunionese had never before experienced chikungunya, and many of them interpreted the disease in relation to the malaria outbreaks of the 1950s. The location of the breeding grounds of the *Aedes* mosquitos in the *jardin creole*—the garden adjoining, and much cherished by, traditional Réunionese households—as well as local people's experiences with previous public health interventions against vector-borne diseases, contributed to the popular stigmatization of chikungunya. Islanders viewed mosquitos as familiar entities in local gardens and French public health agents carrying out mosquito control as trespassers invading the intimacy of the private garden. Moreover, elderly Réunionese associated chikungunya public health interventions with their past experiences of governmental malaria control measures that targeted unsanitary households. This led to the stigmatization of chikungunya sufferers as people whose neglect of domestic hygiene resulted in the creation of mosquito breeding grounds. Stigmatization, in turn, led many Réunionese to reject the idea that chikungunya was a vector-borne disease. At the beginning of 2006, when the chikungunya epidemic peaked—over 25,000 new cases being registered in a single week—there was a widespread belief that the disease was the result of a medical or military experiment, the workings of a chemical plant, or even of a terrorist attack. This heightened public opposition to official health agents. Thus, Jansen argues, public health interventions may have contributed to an increase rather than a decrease in the impact of chikungunya on Réunion during the 2005–2007 epidemic.

In the final contribution to this volume, Eva-Maria Knoll draws an arc from a legendary historical malady to the current impact of malaria-causing parasites in the Republic of Maldives. This small archipelago, lying at the crossroads of the IOW maritime trade routes, has the world's highest rate of a care-intensive inherited blood disorder (beta-thalassaemia). With its current focus on genetic risk, public health discourse in the Maldives is turning the tropical paradise islands into a thalassaemia risk-alert social world. It does not, however, offer a causative explanation for the archipelago's exceptional thalassaemia burden.

Knoll investigates European and Arab historical reports of the "Maldive fever", a malady that most scholars assume to have been malaria. These reports date back about six centuries, starting with Muslim traveller Ibn Battuta in the fourteenth century, followed by castaway François Pyrard de Laval, East India Company surgeon David Campbell, Archaeological Commissioner of Ceylon H.C.P. Bell, and, in the late nineteenth century, collector and explorer Carl Wilhelm Rosset. By the mid-twentieth century, the small Maldives island habitats were recognized as mixed infection zones where the local population had to struggle with three different malaria parasites, the health impacts of which reinforced historical eyewitness reports of inescapable and deadly fever. WHO malaria control activities started comparatively late, in the mid-1960s, but were highly successful. By the mid-1970s, the deadly malaria parasites had been eradicated from the islands and soon thereafter were also eliminated from the collective memory of the Maldivian population.

Disease Dispersion: Tracing Dynamic Biosocial Phenomena

This volume comprises contributions on disease dispersion and impact from a number of academic and scientific angles, including history, social and medical anthropology, archaeology, epidemiology, and palaeopathology. These show that disease-causing agents often took advantage of Anthropocene environmental transformations such as deforestation, monoculture, and irrigation systems, and that they adapted to new environmental circumstances.

Diseases are highly *dynamic biosocial phenomena*, which often makes it difficult to identify them retrospectively and trace them in the historical record (Cooter and Stein 2013; Cunningham 2002; Arrizabalaga 2002; Wilson 2000). DNA analysis demonstrates that some pathogens in the IOW have relatively stable genomes. This is the case, for example, with *Mycobacterium leprae* (which causes leprosy), which has changed little over at least the last 1000 years. Other pathogens, however, may alter considerably over time. Their biological dynamic allows diseases to emerge unexpectedly; to strike, retreat, and re-emerge; and to affect a specific locality, where they might become endemic, or expand in the form of epidemics and even pandemics. Evolutionary mutations enable diseases to change genetically, offering them the potential to transmute, move, and adapt to new environments and circumstances (Alpers, this volume).

Here, we consider diseases to be both "real entities" and "thought entities" (Cunningham 2002, 15)—both biomedical realities and the results of historically produced knowledge, analysed according to either naturalist-realist or historical-conceptualist perspectives (Wilson 2000). The naturalist-realist approach uses modern concepts of disease and biomedical diagnosis to examine what historian Piers Mitchell (2011) calls "social diagnosis" made by people in the past. Thus some scholars identify malaria with certain historical descriptions of "fever". By contrast, the historical-conceptualist approach acknowledges that a general understanding of the nature of a specific disease is embedded in the "thought collective" of a particular time and society (Fleck 1979). Aetiological and epistemological theorizing, by consequence, may change over time. Such changes in the descriptive and diagnostic categories of disease are sometimes difficult to discern (Strahorn, this volume).

Most contemporary medical historians adopt a mixed approach, combining the methodologies used in the humanities with those employed in biomedicine, molecular biology, and genetics, in order to shed new light on historic categories and descriptions. The chikungunya virus, for example, was identified scientifically in 1952 and thus distinguished from dengue and the broader fever category (Alpers, this volume). Since "fever" is a comprehensive disease category used extensively in historic descriptions, some aetiological confusion between dengue, chikungunya, malaria, and other variants of tropical fever characterizes the medical history of the IOW (Kuno 2015, Knoll, this volume). Genomic studies of tuberculosis (TB), by contrast, have expanded the TB category to the *Mycobacterium tuberculosis* complex (MTBC), comprising several species of the causative organism. Furthermore, they have shown that tuberculosis is much younger than hitherto assumed. Rather than representing a prehistoric disease originating from "out-of-Africa" migrations, tuberculosis dates back only 4000 to 4500 years (Green and Jones, this volume).

Sequencing ancient DNA (aDNA), retrieved from bones and dental remains, is the most recent development in the methodological toolkit for the reconstruction of human ailments and disease in history (Hagelberg et al. 2015; Strahorn, this volume). Though technically challenging, and of limited utility in the reconstruction of infectious and genetic disease history, aDNA results provide new kind of data. This invites, if not obliges, a reassessment of previous historical data and resulting insights and arguments that were largely based on European text sources.

This volume invites readers to consider disease in history from both a life science and social science/humanities perspective. Such an approach takes pathogen and gene pools as well as disease perception and experience into account. However, the poor preservation of human skeletons, especially in the warm and humid climates of the IOW, imposes a serious limitation on the aDNA revolution. Thus Green and Jones (this volume) argue that no pathogen aDNA samples of any of the five major infectious diseases of the IOW—malaria, tuberculosis, leprosy, smallpox, and plague—have yet been retrieved from this macro-region. Finally, the disease history of the IOW also reveals a wide range of perceptions about disease aetiologies and epidemiologies. For example, some IOW communities believed that disease was linked to supernatural forces or blamed it on social "others"—such as Indian "coolies", immigrants, slaves, the crews of visiting ships, pilgrims, or colonizers.

In sum, this collection focuses on disease dispersion across time, space, and various cultural settings rather than on, for example, the large field of culture-specific diseases captured as culture-bound syndromes or folk illnesses. It will hopefully stimulate scholars to engage in other, interdisciplinary studies, of the biosocial dispersion of disease in the IOW, past and present.

Acknowledgements The conference from which this book emerged took place in September 2016 at McGill University in Montreal and was chiefly funded by the Social Sciences and Humanities Research Council of Canada (SSHRC). We also thank the Max Planck Fellow Group "Connectivity in Motion: Port Cities of the Indian Ocean" at the Max Planck Institute for Social Anthropology, Halle, Germany, under the directorship of Max Planck Fellow Burkhard Schnepel, for his financial assistance towards the conference and for his preparatory work on the resulting volume. We are grateful to the conference contributors and participants for lively discussions and substantial suggestions. We also thank the two anonymous reviewer for their valuable remarks and, last but not least, Tyler Yank for her copyediting work.

References

Alpers, Edward A. 2000. Indian Ocean Africa: The Island Factor. *Emergences: Journal for the Study of Media & Composite Cultures* 10 (2): 373–386. https://doi.org/10.1080/713665815.

———. 2014. *The Indian Ocean in World History*. Oxford and New York: Oxford University Press.

Arnold, David. 1991. The Indian Ocean as a Disease Zone, 1500–1950. *South Asia: Journal of South Asian Studies* 14 (2): 1–21. https://doi.org/10.1080/00856409108723152.

Arrizabalaga, Jon. 2002. Problematizing Retrospective Diagnosis in the History of Disease. *Asclepio* 54 (1): 51–70.

Bos, Kirsten I., et al. 2016. Eighteenth Century *Yersinia pestis* Genomes Reveal the Long-term Persistence of an Historical Plague Focus. *eLife* 5: e12994. https://doi.org/10.7554/eLife.12994.

Campbell, Gwyn. 2014. The Question of Slavery in the Indian Ocean World History. In *The Indian Ocean. Oceanic Connections and the Creation of New Societies*, ed. Abdul Sheriff and Engseng Ho, 123–149. London: Hurst & Company.

———, ed. 2018. *Bondage and the Environment in the Indian Ocean World*. Palgrave Series in Indian Ocean Studies. Basingstoke and New York: Palgrave Macmillan.

———. 2019. *Africa and the Indian Ocean World from Early Times to 1900*. Cambridge: Cambridge University Press.

Cliff, Andrew David, Peter Haggett, and Matthew Smallman-Raynor. 2000. *Island Epidemics*. Oxford: Oxford University Press.

Cohn, Samuel K., and Lawrence T. Weaver. 2006. The Black Death and AIDS: CCR5-Δ32 in Genetics and History. *QJM: An International Journal of Medicine* 99 (8): 497–503. https://doi.org/10.1093/qjmed/hcl076.

Cooter, Roger, and Claudia Stein. 2013. *Writing History in the Age of Biomedicine*. New Haven, CT: Yale University Press.

Crosby, Alfred W. 1972. *The Columbian Exchange: The Biological and Cultural Consequences of 1492*. Westport: Greenwood Publishing Company.

Cunningham, Andrew. 2002. Identifying Disease in the Past: Cutting the Gordian Knot. *Asclepio* 54: 13–34.

Falola, Toyin, R. Joseph Parrott, and Danielle Porter Sanchez. 2019. Introduction: Arbiters and Witnesses of Change. Contextualizing Conversations on African Islands. In *African Islands: Leading Edges of Empire and Globalization*, ed. Falola Toyin, R. Joseph Parrott, and Danielle Porter Sanchez, 1–35. Rochester, NY: University of Rochester Press.

Fleck, Ludwik. 1979 [1935]. *Genesis and Development of a Scientific Fact*. Ed. T.J. Trenn and R. Merton and Trans. F. Bradley and T.J. Trenn. Chicago and London: The University of Chicago Press.

Foucault, Michel. 1973. *The Birth of the Clinic: Archaeology of Medical Perception*. London: Routledge.

———. 2011. *The Government of Self and Others: Lectures at the College de France, 1982–1983*. Paperback ed. Basingstoke: Palgrave Macmillan.

Hagelberg, Erika, Michael Hofreiter, and Christine Keyser. 2015. *Ancient DNA: The First Three Decades. Philosophical Transactions of the Royal Society B: Biological Sciences* 370: 1–6. https://doi.org/10.1098/rstb.2013.0371.

Issa, Amina A. 2006. Dhows and Epidemics in the Indian Ocean Ports. *Ziff Journal* 4: 63–70.

Knoll, Eva-Maria. 2017. Archipelagic Genes: Medical Travel as a Creative Response to Limitations and Remoteness in the Maldives. *Asia Pacific Viewpoint* 58 (2): 148–161. Special Issue: International Medical Travel and the Politics of Transnational Mobility in Asia, edited by Heng Leng Chee, Andrea Whittaker and Brenda S.A. Yeoh. https://doi.org/10.1111/apv.12157.

Kuno, Goro. 2015. A Re-Examination of the History of Etiologic Confusion between Dengue and Chikungunya. *PLOS Neglected Tropical Diseases* 9 (11): e0004101. https://doi.org/10.1371/journal.pntd.0004101.

Meunier, Yann A. 2014. *Tropical Diseases. A Practical Guide for Medical Practitioners and Students.* Oxford and New York: Oxford University Press.

Mitchell, Piers D. 2011. Retrospective Diagnosis and the Use of Historical Texts for Investigating Disease in the Past. *International Journal of Paleopathology* 1 (2): 81–88. https://doi.org/10.1016/j.ijpp.2011.04.002.

Mitchell, Peter. 2016. *Under Pressure: Steamships, Global Power and Communications, and the East India Company—Part 1.* Snapshots of Empire website, University of Sussex. http://blogs.sussex.ac.uk/snapshotsofempire/2016/03/07/under-pressure-steamships-global-power-and-communications-and-the-east-india-company-part-1/. Accessed 17 Jan 2019.

Pearson, Michael N. 2003. *The Indian Ocean.* London and New York: Routledge.

———., ed. 2015. *Trade, Circulation, and Flow in the Indian Ocean World.* Basingstoke: Palgrave Macmillan.

Reid, Anthony. 2018. Recognising Global Interdependence Through Disasters. In *Crossing Borders. Governing Environmental Disasters in a Global Urban Age in Asia and the Pacific*, ed. Michelle Ann Miller, Michael Douglass, and Matthias Garschagen, 21–40. Singapore: Springer.

Schnepel, Burkhard, and Edward A. Alpers, eds. 2017. *Connectivity in Motion. Island Hubs in the Indian Ocean World.* Basingstoke: Palgrave Macmillan.

Sheriff, Abdul. 2010. *Dhow Cultures of the Indian Ocean: Cosmopolitanism, Commerce and Islam.* London: C. Hurst & Company.

Sheriff, Abdul, and Engseng Ho. 2014. *The Indian Ocean. Oceanic Connections and the Creation of New Societies.* London: Hurst & Company.

Sussman, George D. 2011. Was the Black Death in India and China? *Bulletin of the History of Medicine* 85 (3): 319–355.

Weaver, Scott C., and Marc Lecuit. 2015. Chikungunya Virus and the Global Spread of a Mosquito-Borne Disease. *The New England Journal of Medicine* 372: 1231–1239. https://doi.org/10.1056/NEJMra1406035.

Wilson, Adrian. 2000. On the History of Disease Concepts: The Case of Pleurisy. *History of Science* 38 (3): 271–319. http://eprints.whiterose.ac.uk/3330/.
Winterbottom, Anna, and Facil Tesfaye, eds. 2016. *Histories of Medicine and Healing in the Indian Ocean World.* 2 Vols. Palgrave Series in Indian Ocean Studies. Basingstoke and New York: Palgrave Macmillan.
Zeller, H.G. 1990. Dengue, Arbovirus and Migrations in the Indian Ocean. *Bulletin de la Societe de Pathologie Exotique* 91 (1): 56–60.

CHAPTER 2

The Evolution and Spread of Major Human Diseases in the Indian Ocean World

Monica H. Green and Lori Jones

Over the past two decades, developments in molecular genetics have allowed the systematic reconstruction of the evolutionary histories of pathogenic organisms, including those responsible for malaria, tuberculosis, leprosy, smallpox, and plague. Such reconstructions have significantly increased knowledge about the history of these global infectious diseases. They have also forced a reconsideration of the places where these diseases originated, how they evolved, and how they spread around the world. This chapter argues that although the Indian Ocean World (IOW) has not yet factored prominently in global health history, and has yet to yield any samples of pathogen ancient DNA (aDNA), it is nevertheless possible to

M. H. Green (✉)
Independent Scholar, Phoenix, AZ, USA
e-mail: mhgreen@asu.edu

L. Jones (✉)
Carleton University, Ottawa, ON, Canada

University of Ottawa, Ottawa, ON, Canada
e-mail: ljone041@uottawa.ca

G. Campbell, E.-M. Knoll (eds.), *Disease Dispersion and Impact in the Indian Ocean World*, Palgrave Series in Indian Ocean World Studies, https://doi.org/10.1007/978-3-030-36264-5_2

begin to reconstruct the pre-1500 histories and conceivable spread of each of these major human diseases across this region of the world. Starting from the fact that all five diseases were endemic in the IOW in the twentieth century, this chapter asks three fundamental questions: When did they arrive in the IOW? From where did they arrive? and How might connectivities within and through the IOW have contributed to their global spread?

The field of the history of disease, or historical epidemiology, is in the midst of a sea change. Modern scientific understanding of infectious diseases experienced a world-altering shift in the late nineteenth century with the full development of the germ theory of disease. The advent of laboratory-based analysis allowed scientists to both identify and disrupt the processes by which microbial pathogens replicated, spread, and affected their hosts. However, the laboratory-based model of identifying pathogens, and proving their causative role in human disease, had little to contribute to the *history* of disease. Laboratories could only work on living cells or on cells (or viruses) that had been collected under controlled conditions. It was possible to *infer* the presence of particular diseases in the pre-laboratory past, but impossible to *prove* their historical existence according to the accepted standards of modern microbiology.

That evidentiary barrier has now been breached. With the institution of molecular methods of studying microbes in the 1980s came the possibility of studying not simply whole, intact bacteria, viruses, and other microscopic organisms but also their molecular components. Starting in the 1990s, researchers began to use molecular methods to prove the presence of particular pathogens in historical samples, such as mummified tissues, skeletons, and tooth pulp. This new field of palaeogenetics (also called *archaeogenetics*) builds on bioarchaeological investigations that look both macroscopically and microscopically at the evidence of human remains. It is revolutionizing knowledge about the histories of some of the world's most important infectious diseases.

As late as 2015, a historian of disease in the modern era, Mark Harrison, noted that "there are no Asian counterparts to the Atlantic studies of Alfred W. Crosby and John R. McNeill"—a reference to their pioneering works on the Columbian Exchange, that great transfer of flora, fauna, and microbiota between Africa, Europe, and the Americas during the peak period of European colonialism and transatlantic slavery (Harrison 2015, 650).[1] This gap has persisted despite the existence of considerable

[1] Harrison refers to Crosby (1972, 1986) and McNeill (2010). One could add to the list of pioneering studies on the history of disease in the Atlantic the works of the Africanist Curtin (1961, 1968).

speculation about the history of infectious diseases in the IOW and the availability of written primary sources that appear to discuss them but that have not been adequately studied. A marked Eurocentrism has characterized the history of infectious diseases in general, with even a recent survey of "Plague and Lethal Epidemics in the Pre-Industrial World" making no mention of the IOW (Alfani and Murphy 2017). Historians writing IOW history have tended to emphasize post-1500 European sources, very few of which have any validity for the pre-1500 IOW. David Arnold's otherwise pioneering essay, "The Indian Ocean as a Disease Zone, 1500–1950", for example, relied largely on narratives that were constructed from European accounts (Arnold 1991). Historiography based on local IOW sources does exist, as do many claims about the "antiquity" of various infectious diseases in this region, including smallpox, cholera, plague, and malaria. However, such works tend to draw selectively from references in Chinese, Ayurvedic, and other IOW texts that have often been poorly translated and/or edited, without the necessary rigour in dating or assessing the semantic weight of disease terminology to circumvent the more egregious problems of retrospective diagnosis (Meulenbeld 1999–2002; Wujastyk et al. 2013; Green 2014, 51–53). More fundamentally, the history of disease in the IOW lacks the framing chronology that it has in the Atlantic Basin, where a clear pre- and post-c.1500 divide centred on European exploration, settlement, and eventually colonialism defines its basic "Columbian Exchange" narrative. In 1976, world historian William H. McNeill incorporated South Asia (although not the East African littoral) into his conceptualization of regions affected by what he called "civilized disease pools" (McNeill 1976, 94). For McNeill, the interlinking of China, India, Mesopotamia, and the Roman Empire via sea trade and caravan travel marked a transition in the history of the development and spread of human infectious diseases. This notion of an historical "age of pandemics" has recently been revived by Kyle Harper to explain the role played by infectious disease in the decline of the Roman Empire (Harper 2017). In the present account, we do not claim to have defined a new chronological framework since, as will be seen, the five diseases we examine likely had very different temporal arcs in terms of their arrival in, and dissemination through, the IOW, spanning a few hundred to perhaps tens of thousands of years. We do, however, agree with earlier historians that the active maritime and trading culture that linked the IOW for well over 2000 years sets the context for the history of these infectious diseases even if, as will be seen in the case of plague, other geographical determinants may have played a role as well. In short, the disease history ideas and questions that Crosby, McNeill, Arnold, and others offered years ago can now be reframed by additional insights and shown to lead to something really new.

The most revolutionary aspect of the new historical epidemiology is that it can both draw on narrative and documentary sources traditionally used by historians and move beyond them. No longer must histories of diseases like leprosy, plague, or tuberculosis be limited to those societies, or those literate levels of a given society, with written traditions of medicine or public health. Nor need they be constrained by systems of understanding and describing disease that have been, by definition, different from currently used biomedical categories. Obviously, even a material history of disease is reliant on the available evidence: it is not a coincidence that the majority of works on palaeogenetics published thus far has relied on European samples (Green 2018a). Nevertheless—and this is important—because palaeogenetics is premised on an understanding of pathogens as organisms subject to evolutionary forces, work on both historical and modern samples is equally revealing. Put simply, anything that exists now had ancestors. It is possible to infer what those ancestors might have looked like from their modern descendants. Moreover, if scientists are fortunate enough to also retrieve "fossils" of those ancestors, it is possible to better understand the entire family tree.

Even in advance of any major palaeogenetic discoveries in the region, enough data is currently available that it is possible to begin to reconstruct the histories of the IOW's major infectious diseases. This work takes its power from its global framework. For globally distributed infectious diseases, inferences about the genetic character of a disease organism in one part of the world can be used to make inferences about it elsewhere. Indeed, that is the great and unique contribution that palaeogenetics makes to the history of medicine. Not only can it confirm the identity of particular pathogens in certain times and places, but it also provides the means to track the paths along which those diseases spread. It allows historical epidemiologists to do what regular epidemiologists do: determine which strains are spreading (have spread), where, and amongst which populations. These results have the potential to overturn long-established narratives that have hitherto been based solely on written documentary sources or on a handful of cases described by palaeopathology (Green 2017, 494–520). One of these narratives is the idea that European colonialism was a key turning point in IOW disease history, suggesting, for example, that tuberculosis (TB) was not a problem in India until the 1840s

(Harrison 2015). However, that assertion is belied by lineages of TB, the current distributions of which are heavily concentrated in India where their appearance probably preceded the European arrival by hundreds if not thousands of years.

The following analysis, therefore, adopts an evolutionary perspective on the connectivity of infectious diseases in the IOW. The "story" embedded in the genomes of pathogenic organisms tells of origins and wanderings and of divergent branches which, because of distinctive genetic markings, can be separated and studied as distinct historical events. The history of infectious diseases in the IOW can now be explored.

Infectious Diseases in the Indian Ocean Basin

There is ample evidence that malaria, tuberculosis, leprosy, and plague are all highly endemic around the Indian Ocean basin and that smallpox was similarly endemic until the 1970s. Moreover, various types of evidence suggest the *plausible* or *possible* presence of each of these diseases in the region prior to 1500. We privilege here the work of genetics which, because of the imperative of evolutionary analysis, forces connections to be made with a deeper past. In fleshing out these narratives in the future, however, all historicist fields must contribute. This includes bioarchaeology—the contextualized study of all physical indicators of lifestyle, diet, stress, and disease, of individuals and of larger communities (Robbins Schug and Blevins 2016; Baker and Agarwal 2017; Clark et al. 2017)—which is transforming knowledge of, for example, the connections between East Africa and other maritime cultures (Hoogervorst 2016; Horton 2017; Wynne-Jones and LaViolette 2017). Similarly, the ability to interrogate the written record (i.e., documentary sources) for IOW societies is growing apace, with new digital technologies making the work of editing and analysing textual materials increasingly rigorous.[2] In defining the geography of the IOW, this chapter includes all landmasses abutting the region. Only Australia has thus far yielded scant information on its pre-modern disease history. The very early peopling, and subsequent isolation, of that landmass may have shielded it from many if not all of the diseases studied here up to modern times.

[2] For example, work on the Cairo Genizah, the repository of records from the eleventh to nineteenth centuries, documents the vast IOW trade networks of Jewish families based in Cairo, Yemen, and Mangalore (Cambridge University Library 2017).

Malaria: Maps depicting the modern distribution of the *Plasmodium vivax* and *Plasmodium falciparum* strains of malaria show a stark contrast. As falciparum appears to be concentrated in Africa and in those areas of the world that received the highest numbers of Africans during the early modern slave trade, while vivax seems to have an Asian centre of distribution (Gething et al. 2011, 2012), it was until recently assumed that the former was the older strain and that the latter may have originated in Asia. Recent genetics work, however, has clearly demonstrated that vivax is both much older than falciparum and also is of African genesis (Loy et al. 2017).

Primate studies demonstrate that in Africa a vivax progenitor probably circulated amongst primate species and that humans were just one of several regular hosts. When early humans migrated out of Africa and into regions that lacked other large primate species, there were no longer multiple primates through which the organism could cycle. The disease thus adapted to become a specialized disease of humans (Liu et al. 2010, 2014; Larremore et al. 2015; Webb 2017). It is not yet known how early vivax malaria moved out of Africa, but recent studies suggest that it is necessary to consider the very real possibility that it moved into the IOW with the predecessors of modern humans (i.e., Neanderthals and Denisovans) and only there did it become a distinctly human disease (Houldcroft and Underdown 2016; Meyer et al. 2016).

Falciparum malaria, by contrast, may have originated as a human disease in Africa as recently as 10,000 years ago through a sudden and unique species transference from gorillas (Sundararaman et al. 2016; Loy et al. 2017). Studies of ancient DNA and of human immune profiles suggest the presence of falciparum malaria in ancient civilizations around the Mediterranean, while medical textual evidence of malaria infection suggests widespread familiarity with the disease in the Mediterranean world and in Asia as far east as China by 1000 BCE (Webb 2009; Viganó et al. 2017; Marciniak et al. 2018).

Malaria's "out-of-Africa" progression to the east seems to have stopped before it reached Australia and the Pacific Islands (Buckley 2006). Moreover, it did not accompany early human migration to the Americas. The subsequent, and much later, spread of malaria beyond the IOW was largely the result of human action. Civil engineering and urban design projects created vector-friendly micro-environments in the Roman Empire (Ziegler 2016; Marciniak et al. 2018), and Roman colonialism spread malaria across transalpine Europe (Newfield 2017). Again, falciparum was carried in the bodies of slaves and slave traders across the Atlantic to the

Americas where the use of specific farming technologies in sugar and cotton plantations enabled the adaptation of the disease to its new surroundings (McNeill 2010). In the case of malaria, then, early human migration established the disease around the IOW. Later human migration, aided by technological advances, most especially rice cultivation, which involved the use of flooded fields that were amenable to mosquito production, helped it to spread further afield (King et al. 2017).

Tuberculosis (TB): Tuberculosis is another disease the narrative history of which has recently been turned on its head by genomic studies. Current estimates suggest that, even after aggressive interventions in the early twenty-first century, close to one-fourth of all people alive today carry TB in their bodies. For Southeast Asia, prevalence may run as high as 30 per cent, the highest for any region of the world (Houben and Dodd 2016). In genetics, tuberculosis is now referred to as the MTBC (*Mycobacterium tuberculosis* complex), since it comprises more than one species of the organism. The traditional narrative of TB's history, which is now being challenged by palaeogenetics, connected it with the main phase of human migration out of Africa. It suggested that the disease developed at least 70,000 years before the present (BP) in Africa, where it split into two branches, one associated with West Africa and the other with the eastern part of the continent. As humans migrated out of Africa, they carried the disease not only to Europe and Asia but also, prior to the Columbian Exchange, via the Bering land bridge to the Americas. This dispersion of humans and their bacteria formed the seven main TB lineages that are found in the world today (Gagneux 2012; Luo et al. 2014).

The latest palaeogenetic interpretations, in contrast, profoundly challenge the chronology and at least part of the geographic trajectory of the traditional narrative. In 2014, the complete sequencing of 1000-year-old tuberculosis aDNA from South America, combined with an eighteenth-century sample from Europe, suggested that TB's history as a human disease was much more shallow. That study and others since have suggested that TB is probably less than 6000 years old, perhaps closer to about 4000 to 4500 years—meaning that all the dispersal and lineage formation occurred in a much more limited time period (Bos et al. 2014, 496; Comas et al. 2015; Kay et al. 2015; Honap et al. 2017; O'Neill et al. 2017). This 4000- to 4500-year window for TB's history has been challenged by palaeopathologists, who believe that distinctive skeletal signs of TB infections can be found at least 3000 or 4000 years earlier (Sparacello et al. 2016). Even so, the drastically foreshortened timeframe for TB's entire history in humans rules out a simple in-and-out-of-Africa and Beringia-passage-to-America narrative simply on the issue of dating: a pathogen not older than 6000 or even 8000 years cannot have been involved in a human migration that took place 17,000 to 20,000 BP (Green 2017, 498–502).

Africa is the only continent where all seven human-adapted MTBC lineages are currently found (Gagneux 2018). The wide diversity of TB strains found in East Africa in particular, and the decreasing diversity of strains found with increasing distance from Africa, has led to the recent suggestion that the Horn of Africa, on the north-western edge of the IOW, was the original birthplace of MTBC as a human disease. After an early split that separated the West African strains (the two lineages *M. africanum*) from the rest, all other TB lineages in the world today developed from a common ancestor that had its origins in Northeast Africa (Coscolla and Gagneux 2014; O'Neill et al. 2017). What epidemiological circumstances allowed the disease to emerge are unclear. Most mycobacteria are saprophytes, that is, soil-based organisms primarily involved in breaking down decaying matter. How, then, did the organism become a human-obligate pathogen that relied on respiratory transmission? New work on a related species that shares a common ancestor with the MTBC, *Mycobacterium canettii*, complicates that question further (Supply and Brosch 2017). Specifically, our current knowledge cannot account for the adaptations that would have been needed to turn the organism into a "professional pathogen": that is, one that had no other lifecycle save as an infectious agent, which needed to create a diseased state in its host to be effectively transmitted (Gagneux 2018). Whatever the mechanisms of adaptation were, once MTBC did become a "professional pathogen", we must assume that human migration and trade played a significant role in the subsequent spread of the disease.[3]

No detailed study of the long-term history of TB in the IOW has yet been published. Although TB can produce distinctive skeletal lesions if it advances to a certain stage of infection outside the lungs (Roberts and Buikstra 2003), only limited bioarchaeological evidence for TB's presence

[3] We assume for now that the IOW MTBC lineages were obligate human pathogens, transmitted solely from human host to human host. In the Americas, by contrast, seals brought a zoonotically transmitted strain to South America at least as early as 1000 CE; it then spread via human-to-human transmission for several hundred years before being superseded by strains brought to the New World by Europeans (Bos et al. 2014; Honap et al. 2017; Green 2017, 498–502).

in the IOW has appeared.[4] TB seems to have been present, for example, in the early second millennium BCE Indus River Valley, which fits within the shallower chronology that palaeogeneticists have recently proposed (Robbins Schug et al. 2013). For now, though, the history of TB in the IOW comes from genetics and the growing global understanding of TB's evolutionary history. This shows that there was not one but three major lineages of TB connected closely to the IOW (Fig. 2.1). The first, the

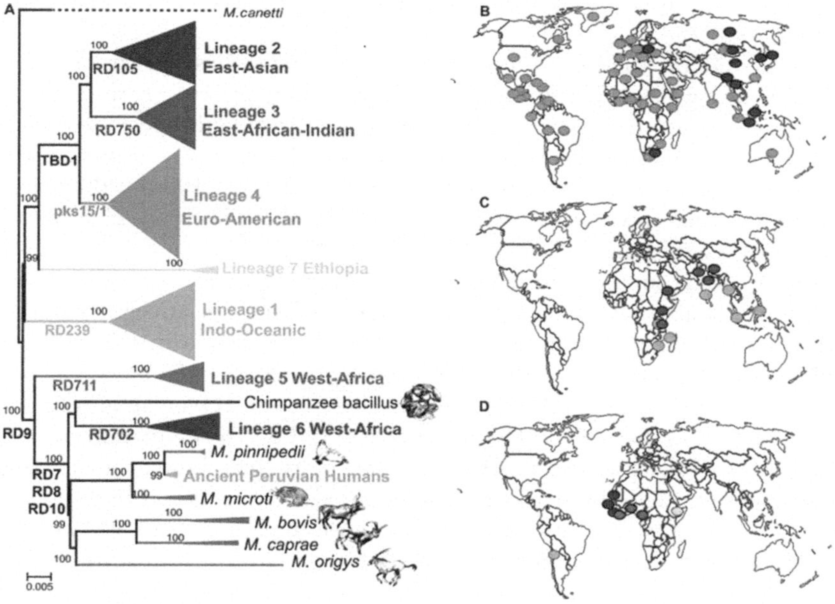

Fig. 2.1 A phylogenetic tree showing the relationships of the seven MTBC lineages. Maps B–D show, respectively, the global distribution of Lineages 2 and 4, Lineages 3 and 1, and Lineages 5, 6, 7, and (in South America) an extinct Ancient Peruvian lineage. (Source: Coscolla and Gagneux (2014, 434). Initially produced under a Creative Commons Licence (Attribution-NonCommercial-ShareAlike 3.0 Unported [CC BY-NC-SA 3.0]).)

[4] Roberts and Buikstra (2003, 130–131) note that no bioarchaeological data for TB's premodern history had yet been found for the following IOW countries: all of sub-Saharan Africa, Burma, Indonesia, Malaysia, Vietnam, Laos, New Zealand, Australia, India, Bangladesh, Nepal, Pakistan, Saudi Arabia, United Arab Emirates, Oman, Kuwait, Bahrain, Yemen, and Afghanistan.

Indo-Oceanic lineage (Lineage 1), is the most genetically diverse and probably the oldest of the extant lineages. Its various sublineages are all very different, which suggests that there was a long period of time over which each was able to acquire multiple mutations. A recent study has estimated that Lineage 1 diverged from its most recent common ancestor (MRCA) with the main MTBC between 2765 and 5104 years ago. Extant sublineages seem to have begun diverging between 1471 and 2286 years ago. The current geographic range of this lineage is concentrated around the rim of the IOW, from Afghanistan to Vietnam (Coscolla and Gagneux 2014, 433).[5]

A second TB lineage is almost entirely confined to East Africa. Lineage 7, also known as the Ethiopia lineage, branched off very early from a MRCA that it shared with the other East African lineages, probably between 2659 and 4904 years ago. However, divergence within the lineage itself seems to have happened only within the past 400 to 900 years, suggesting that this lineage survived within a small, isolated population for much of its existence (Comas et al. 2015). Finally, Lineage 3, the East African-Indian lineage, probably shared a common ancestor with other TB lineages about 1653 to 3113 years ago, making it relatively young in genetic terms. Its sublineage diversity dates from approximately 877 to 1717 years ago (Bos et al. 2014; Coscolla and Gagneux 2014).

The retrieval of TB aDNA from anywhere in the IOW will automatically add greater specificity to this narrative. If TB's dissemination out of East Africa into the wider IOW was due to regular human-to-human transmission, then the shallower timeframe recently proposed by genetics puts connectivity through regional migration and trade (i.e., across the IOW) at the forefront of the epidemiological patterns seen in current distributions, more so than the much older large-scale human migration out of Africa previously assumed in the longer timeframe. Even now, without aDNA, the new genetics shows the importance of understanding different strains of the disease as constituting, in effect, epidemiological strata that can be connected to different historical periods and events. The collected evidence from genetics gives no reason to believe, for example, that TB

[5]The 43 Lineage 1 samples used for the major study that re-dated the MTBC (Bos et al. 2014) came from patients born in Afghanistan, Burkina Faso, Burma, Cambodia, China, Comoro Islands, Ethiopia, Ghana, India, Laos, Nepal, the Philippines, Serbia, Singapore, Sri Lanka, Thailand, and Vietnam. On the within-lineage diversity of Lineage 1, see Coscolla and Gagneux (2014, 433).

was a new importation into any part of the IOW during the colonial period, as some historians relying on written documentary sources have suggested (Harrison and Worboys 1997; Harrison 2015).

Even if the bioarchaeological record is still thin, the existence of two old lineages of TB, the current distribution of which is heavily concentrated in India, points to the presence of the disease in South Asia long before the arrival of European colonialists. Lineage 4, of which some ten sublineages have recently been identified, has been dubbed the Euro-American lineage, because it seems most strongly characteristic of Europe and modern North America. Some early strains of Lineage 4 may have reached Ethiopia 3000 to 4000 years ago, perhaps from Central Eurasia (Comas et al. 2015), but its presence in Europe has been presumed to date to at least the period of Roman expansion (Kay et al. 2015). The Roman Empire then likely played a major role in creating the main pattern of disease dissemination that ultimately made this lineage "European". More virulent sublineages of Lineage 4 are quite recent in origin; these have dispersed throughout the world and "swamped" pre-existing local indigenous strains, in effect mirroring almost exactly Europe's global colonialism and erasing evidence of earlier TB distributions. Such seems to have been the fate of the seal-derived strain of TB found in the pre-Columbian Americas. Such "swamping" by a more virulent strain carried by Europeans may also have happened in South Africa (Coscolla and Gagneux 2014; Stucki et al. 2016). In looking at modern maps of TB's distribution around the IOW, therefore, it is important to recognize that European global expansion from c.1500 CE, and the more virulent TB strains that accompanied it, seems to have added a further health burden to this already epidemiologically challenged region of the world.

Leprosy: In the face of effective treatment, the modern map of leprosy distribution differs considerably from a comparable map from the late nineteenth century.[6] Still, even in the early twenty-first century, around 250,000 people in 121 countries are diagnosed with leprosy annually. This figure has held steady despite universal availability of an effective therapeutic regimen since the 1980s (Smith et al. 2017). The identification in 2008 of a second species of leprosy bacillus (*Mycobacterium lepromatosis*) reveals that leprosy has a much more complicated history with the human species than has yet been understood (Singh et al. 2015; Han 2017).

[6] See Thin (1891) for a map of leprosy's late nineteenth-century distribution.

Traditional narratives of leprosy's history have drawn from written documentary sources, to which palaeopathological evidence has been added in recent decades. Although textual references to "leprosy" purportedly date back to about 600 BCE, no persuasive clinical description of the disease can be found before the first centuries BCE and CE. But leprosy can, like TB, cause distinctive lesions in the bones from which a skilled palaeopathologist can diagnose the disease. Currently, the world's oldest known sample of human skeletal remains showing characteristic infection with leprosy is from India and dates back some 4000 years (Robbins et al. 2009). Additional cases have been identified in India from around the turn of the first millennium BCE (Robbins Schug 2016) and in Iron Age Thailand from 300 to 200 BCE (Tayles and Buckley 2004).[7]

The new genetics narrative for leprosy complicates the implications from bioarchaeology and textual references that leprosy originated in the IOW in three ways. First, the two species of leprosy (*Mycobacterium leprae* and *Mycobacterium lepromatosis*), as distinct organisms, are estimated to have diverged from each other nearly 14 million years ago, before the emergence of all hominid species and even the great apes. That means, of course, that leprosy must have developed as a parasite of some other host before being transmitted to humans. When and where that zoonotic transfer happened is as yet unclear, or even whether it has happened multiple times. Second, the six distinct genetic lineages documented in modern clinical samples of the more common type of leprosy, *Mycobacterium leprae*, have an odd geographical distribution, one that does not immediately appear to correlate with known lines of human connectivity in the ancient or mediaeval worlds. The lineage now considered to be the oldest (Lineage 0 or 3K-0) is currently documented in Japan, China, Korea, and New Caledonia. The New Caledonian strain appears in its own clade (subgroup), which diverged from the other 3K-0 strains about 1000 BCE, which may suggest that the Pacific Islands' history with the disease is very old (Benjak et al. 2018). The main lineage that predominates now in India and much of Southeast Asia (Lineage 1), in contrast, is much younger than all other main lineages, having diverged from its most recent common ancestor perhaps as late as the seventh century CE (Fig. 2.2).[8] Third,

[7] For the recently discovered *M. lepromatosis*, we currently know of no distinctive signs in the body's hard tissues that would indicate infection with this organism.

[8] Lineage 0 was discovered after the initial four-lineage and multiple sublineage system was created in 2005 and 2009. Some isolates previously classified as 3K have now been shown to belong to Lineage 0. Lineage 5 was differentiated from Lineage 0 in Schuenemann et al. (2018).

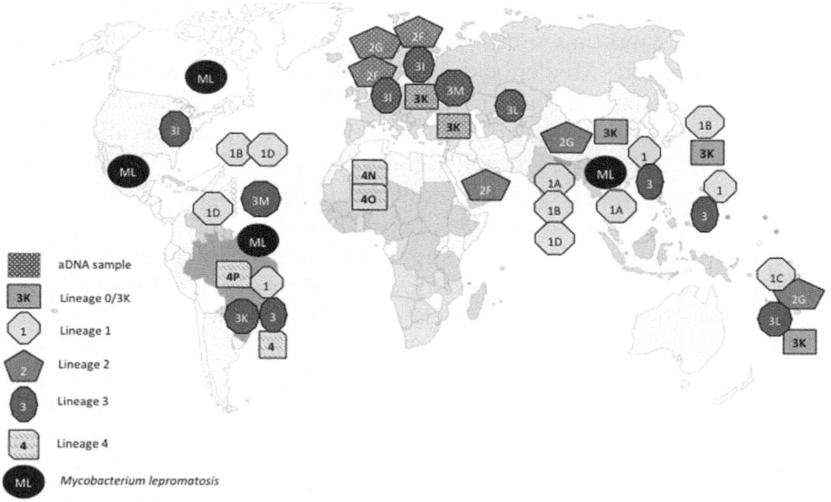

Fig. 2.2 A map showing the global distribution of leprosy lineages Branch 0 to Branch 4, based on both modern DNA and aDNA samples. (Source: Authors)

the sequencing of retrieved aDNA from historical samples, most of which come from Europe, complicate the picture even more. Four of the six known lineages were circulating in Europe by the Middle Ages, including strains closely related to ones now found in East Africa, which diverged from the main lineages of Eurasia sometime around 523 BCE (Schuenemann et al. 2013, 2018; Benjak et al. 2018).

It is unknown whether future genetic discoveries will show whether leprosy in the pre-modern IOW had similar levels of diversity or if that is an artefact of more recent migrations or even zoonotic transmissions. Cultural differences between Europe and the IOW will, in any case, need to be investigated to see if and how local human practices contributed to the dissemination of the disease or to the treatment of those suffering from it. Evidence exists of leprosy in Europe from early in the Common Era, but specialized institutions for those suffering from the disease did not appear with any regularity until the twelfth century. In the Cairo Genizah correspondence relating to Indian Ocean trade, there is only one passing mention of leprosy, in the context of an inescapable fate doled out by God (Goitein and Friedman 2008, 434). In India, written references to a disease with symptoms similar to leprosy date to as early as the first or second century CE (Meulenbeld 1999–2002), and while we know of no formal institutional responses to leprosy, there is clear bioarchaeological evidence of increasing stigmatization and exclusion of persons affected by the disease by at least the first millennium BCE (Robbins Schug 2016). In China, no exclusionary practices can be found until the end of the mediaeval period (Emmerick 1984; Leung 2009).

Leprosy is a very slowly replicating organism: its incubation period can last from two to ten years. It does not cause epidemics, let alone pandemics. Yet, it seems to have been widely distributed in the mediaeval world, possibly as a result of slavery (Mark 2002; Barker 2016). Given that prolonged domestic intimacy seems to be key to leprosy's transmission, it is entirely possible that mediaeval practices of slavery throughout the IOW—which typically involved sexual as well as labour exploitation—rather than other types of connectivity (i.e., local and long-distance trade and communication) are at the heart of the disease's spread. In addition, it has been noted that, unlike much of the Islamicate world or the later trans-Atlantic world, slavery in the Indian Ocean was primarily (albeit not exclusively) local, often the result of personal crises of indebtedness (Campbell 2011). The specific roles played by slavery and other forms of connectivity in the spread of leprosy are, then, one among many questions that an informed historical epidemiology of the IOW must now pursue.

Smallpox: The last known natural case of smallpox appeared in Somalia in 1977, and the disease was declared officially eradicated worldwide in 1980. Instead of incidence, then, a more useful map for smallpox could show the decade in which the disease was eradicated from each country. This would, in turn, reveal that smallpox remained endemic in the IOW longer than it did in many other regions (Roser 2016). It was also probably present in the IOW earlier than any other global region. Recent genetic studies suggest that the independent evolution of gerbilpox, camelpox, and human smallpox from a common cowpox-like ancestor commenced only about 3500 years ago. Given the historical geographical ranges of the naked sole gerbil (the only natural host of taterapox) and camels (the host of camelpox), and the very close genetic relationship between these viruses and the one that affects humans, it is highly likely that they first emerged around the Horn of Africa, following the importation of domesticated camels into East Africa from Arabia (Babkin and Babkina 2012; Zehender et al. 2018). The ancient human population of this region was engaged in crop farming, cattle breeding, and trade, which in turn created both sustained contact between people, rodents, and camels and the conditions necessary for this genetic evolution.

The history of smallpox after this initial cross-species transference is still unknown. Indeed, it is not even clear how long it persisted in Africa. If the disease was smallpox as it became known to modern science, then the short time from infection to resolution (disease interval), and the creation of lifelong immunity in survivors, meant that it either had to keep moving from one community to another or that it established itself as a "childhood disease" among large populations where most of the adults could be assumed to have already been infected. In other words, it is a disease the survival of which requires regularized trade networks, or large urban congregations, or both (Green 2017). It would, then, have found in the IOW an ideal environment.

There are, however, major gaps in the history of the disease. The Hindu goddess Shitala, now closely associated with smallpox, seems only to have coalesced as a figure of worship in Bengal sometime in the tenth to twelfth centuries. In his history of India, the eleventh-century Iranian polymath al-Biruni mentions smallpox as a disease carried on a wind from Sri Lanka. The earliest persuasive textual reference to the disease in China depicts it as having been imported from the west. There would be little point in rehearsing the problems with traditional attempts to divine disease history out of written historical documents, except that phylogenetic analyses, aDNA, or even palaeopathological studies of human remains have not yet been able to provide a definitive scientific narrative (Green 2018a). The worldwide eradication of smallpox in the 1970s eliminated natural infections from which to acquire genetic material, and only recently has a persuasive osteological signature of prior smallpox infections been described. The earliest aDNA thus far retrieved dates only from the early seventeenth century, and the latest computer modelling of the virus's evolutionary history suggests that strains sampled before the disease's eradication may have had a common ancestor no older than ca. 1300 CE; this is a result that even the authors admit does not square with persuasive written accounts that some smallpox-like disease existed well before that date (Duggan et al. 2016; Zehender et al. 2018). Using all methods combined, however, especially textual descriptions of individuals bearing the characteristic lifelong scars of prior infections, might eventually create a substantive history of the disease's impact on this part of the world.

Plague: Plague, the disease caused by the bacterium *Yersinia pestis*, has become the poster child of palaeogenetics studies. It was the first human pathogen fully sequenced from historical remains. It is still at the forefront of the field, having produced the most genome sequences from historical remains (30 at the current count), including several that date to the Bronze Age (3300–1200 BCE). On the basis of hundreds of partially sequenced modern isolates, and, at last count, about 300 completely sequenced genomes, it is now possible not simply to confirm *Y. pestis*'s presence at specific times and places in the past but also to plot out the routes (and potentially the circumstances) of its long-distance transmissions and pandemic effects in the past (Morelli et al. 2010; Cui et al. 2013). While earlier postulations that *Y. pestis* had its ancestral home in the Himalayas (McNeill 1976) have long held sway in both the historical and scientific literature, the current genetics consensus confirms that *Y. pestis* has a northern Eurasian origin. The oldest documented strains (from the third millennium BCE) come from the Caucasus, and during the Bronze Age, they spread from there across the northern Eurasian steppe, all the way to Central Europe (Rasmussen et al. 2015; Andrades Valtueña et al. 2017). Hence, in looking at plague's history in the IOW, the most pressing question is to determine how plague reached the more southerly parts of Eurasia and Africa, which are so distant from its ancestral home. The traditional rat-flea-human causal explanation of plague, which has also long been used to argue that the disease moved along particular routes and by particular modes of transport, likewise needs to be troubled and updated by the addition of other possible vectors and intermediary hosts. Indeed, because the long-distance dissemination of the plague relies very heavily on the intervention (albeit unwitting) of humans, when plague is found far from northern Eurasia, it is imperative to construct plausible explanations of how it moved as far as it did.

There have been three major pandemics of plague in world history: the Justinianic Plague (541–circa 750 CE), the Second Pandemic (initiated by the Black Death in 1346–1353, but lasting until the eighteenth or nineteenth centuries), and the Third Pandemic (1894–circa 1940). Different strains of *Y. pestis* are now associated with each of those pandemics, each with a known genotypic character and a regularized designation that permits analysis of their movements and effects with historical specificity. It is also possible to specify the timing and routes of a variety of other, intermediary strains. In other words, a more finely grained historical understanding of plague now exists than for any other pre-modern infectious disease.

Where the strain is known, it is possible to tease out information about the timing and circumstances of plague's (multiple) arrivals in the IOW. Taking the three pandemics in reverse chronological order, the stories that the modern descendants of *Y. pestis* reveal become clearer.

Plague currently surrounds the IOW; it is found in South and Southeast Asia, India, East Africa, and Madagascar—the last currently having the highest prevalence of the disease in the world (World Health Organization 2008).[9] The highest mortalities of the Third Pandemic occurred in India, which saw an estimated 12.5 million deaths in the early twentieth century. Despite the abundant availability of source materials, there is as yet no synthetic history of the Third Pandemic's effects in the IOW, although some recent studies provide useful insight (Royer 2014).

What really matters with plague is not so much the geography of human cases, but the geography of enzootic foci, the places where the disease persists in rodent populations. Isolates collected in India in the early twentieth century and in subsequent years are located on the same sub-branch of *Y. pestis* as the strain that arrived in the United States around 1900 (Morelli et al. 2010, Supplementary Table 4). These, in turn, have their origin in a strain that emerged from Southeast China (specifically Hong Kong). They are therefore likely to be new imports to the region at the time. They tell us nothing about an earlier history of plague in South or Southeast Asia and in fact reinforce the narrative that the Third Pandemic arose first in Southeast China. Moreover, even strains coming out of Yunnan in the nineteenth century before reaching Hong Kong may have only arrived in that region in the seventeenth century, having originated in a plague reservoir in southern Russia or the Ottoman or Safavid Empires (Cui et al. 2013; Spyrou et al. 2016; Green 2018b). Similarly, genetics tells us clearly that all of the divergent strains now found in Madagascar stem from an introduction of the organism into that island at the end of the nineteenth century (Morelli et al. 2010).

The strains of *Y. pestis* now distributed around the IOW during the Third Pandemic thus form the top layer of the epidemiological archaeology that has been enabled by genetics. Underneath that layer, we can see evidence for the IOW's earlier encounters with the disease. Both India and East Africa had experiences with plague in the early modern period, after-echoes, it seems, of the Black Death. The Second Plague Pandemic is the

[9] Reports are usually collected when there are human outbreaks. The lack of such is no guarantee that the disease does not persist enzootically.

term now used to refer not simply to the infamous mid-fourteenth-century Black Death so gruesomely recounted in sources from both Europe and the Middle East but also to the ensuing waves of plague that afflicted much of Eurasia and North Africa for the following 400 years or more. Current research indicates that the pandemic had two epicentres, one in or near the Caucasus and another in northern China (Hymes 2014; Green 2018b). Both epicentres were offshoots of the major polytomy (divergence) of *Y. pestis* that took place just before the Black Death, which dispersed new strains of plague across the northern Mongol realm. So powerful was the dispersion of *Y. pestis* in this period that fully 80 per cent of all strains of the organism that persist in the world today are descended from this veritable biological explosion.

One of those surviving late mediaeval strains caused an outbreak in India in 2002, affecting 16 people with a 25 per cent mortality rate. That outbreak happened in the very far northwest of the country, in Shimla, Himachal Pradesh, near the point of juncture with northern Pakistan, Afghanistan, and Tajikistan, areas long afflicted with plague (Gupta and Sharma 2007; Mahale et al. 2014). On the principle that whatever exists today must have had ancestors, it is fair to ask whether that strain (which has been typed 2.ANT) was also involved in the reported plague outbreaks in early seventeenth-century India. In the absence of aDNA, there is no way to tell at this point. Similarly, another strain that evolved from the late mediaeval polytomy (2.MED) has been found in modern India, though we have no means to determine when it first arrived in the subcontinent (Kingston et al. 2009).[10] But textual references, coming from both European and Indian sources, support the idea that the plague did arrive in India between the early seventeenth and the mid-nineteenth centuries. A Dutch traveller, Jan Huyghen van Linschoten, who was in Goa in the 1580s, claims that plague "hath never been in India, neither is it known unto the Indians" (van Linschoten 1598/1885). In contrast, Norman Chevers, an English physician and surgeon of the Bengal Medical Service who was in India in 1886 (i.e., before the new Third Pandemic strains would have emerged out of China), readily documents the existence of plague in west India throughout the nineteenth century (Chevers 1886).[11] In between van Linschoten's and Chevers' testimony, we have that

[10] Our thanks to Amy J. Vogler, Northern Arizona University, for information on these studies.

[11] Sussman (2011) is worthwhile in that it collects materials in translation that relate to plague in early modern India. However, our analyses diverge from his by more thoroughly

of the Mughal Emperor Jahangir, who recounts a widespread plague outbreak, centred in the Punjab, in northern India, in 1615 or 1617 (Jahangir 1909–1914). Jahangir describes plague as a new disease at that time. For the moment, therefore, the genetic and written evidence is consilient in pointing towards an early modern arrival of Second Pandemic strains in India. Just how extensively plague had already focalized prior to the arrival of the Third Pandemic strains after 1894 cannot be determined, however, since plague epidemiologists during most of the twentieth century, while recognizing that plague was by then endemic on the subcontinent, did not distinguish between different strains (Sharif 1951). Plague was just "plague".

On the western side of the IOW, in East Africa, the post-Black Death experience with plague was very different. Although it has been suggested that East Africa was infected by a strain coming from India early in the nineteenth century (Sussman 2016), the genetics is quite clear that early modern India and East Africa were likely infected by different strains and, thus, under different circumstances. The strain of *Y. pestis* that now dominates in Kenya, Uganda, the eastern part of the Democratic Republic of the Congo, and Zambia descends in a fairly direct line from the Black Death strain (Green 2014, 34–45, 2018b). Unlike India, where the documented post-Black Death strains belong to the Branch 2 *Yersinia pestis* lineage, the strains in East Africa belong to the Branch 1B lineage, which emerged in the fourteenth century from a plague reservoir in the Caucasus or southern Russia. A recently sequenced fourteenth-century aDNA genome from Bolgar City (Republic of Tatarstan) shows the continued development of that lineage in the region (Spyrou et al. 2016). The strain documented from modern samples in East Africa (called 1.ANT) is an early offshoot of that lineage, earlier even than any strain that made the "back-to-Asia" passage that eventually seeded new strains of Branch 1 in southeast China and spawned the Third Pandemic. It is likely, therefore, that the East African strain and all the other, later "back-to-Asia" Branch 1B strains emerged from a common reservoir in the fifteenth century.

So, by what routes did this Black Death offshoot reach East Africa? Michael Dols, in his still authoritative study of the Black Death in the Middle East, found no evidence for the Persian Gulf's involvement in the early phases of the pandemic and limited evidence for the Red Sea (Dols

drawing from the genetic studies that had already started appearing when he published, which overturn traditional classifications of the different strains of *Y. pestis* and allow the tracking of plague in a decisive way.

1977). Similarly, despite the Ottoman Empire's intermittent involvement with the east coast of Africa (Casale 2010), Nükhet Varlık's masterful study of plague's devastating effects in the early Ottoman Empire finds little to suggest that imperial networks spread plague to its outposts on the edge of the IOW before 1600 (Varlık 2015). While it is possible that plague travelled gradually up the length of the Nile River Valley to its headwaters, an outgrowth of the repeated outbreaks that Lower Egypt suffered in the fourteenth and fifteenth century, the distinctive character of the East African strains instead suggests a route through the Timurid Empire, which seems to have brought a devastating plague outbreak to the Horn of Africa in the fifteenth century (Green 2018b; Derat 2018).[12] Here is a case where having a sample of aDNA from anywhere in the region would do a great deal to clarify the question.

While the currently available evidence is ambiguous about when precisely the IOW was affected by plague in the later Middle Ages and early modern period, for the beginning of the Common Era, the larger question is via what routes it was affected. Ancient DNA samples of Justinianic Plague strains collected thus far from areas north of the Mediterranean (Bavaria, Spain, France, and England) cannot yet help plot out a path back to western China, even though it is clear which Central Asian strains are ancestral to the western pandemic strains (Cui et al. 2013; Wagner et al. 2014; Keller et al. 2019). Written documentary sources demonstrate that the Justinianic Plague ravaged a wide area from at least Persia to the Mediterranean to Ireland between the mid-sixth and mid-eighth centuries after first appearing on the southern shores of the Mediterranean, in Pelusium on the eastern Nile Delta, in July 541 (Little 2007; Mitchell 2014; Harper 2017). Importantly, however, we need to remember that plague is not normally a human disease. Its first appearance in Pelusium thus may be an artefact of human perception, rather than historical and scientific fact.

Did it first arrive in Egypt via the IOW or via the Black Sea and the Mediterranean? At the moment, genetics can provide no help in answering that question, other than telling us that plague had travelled a very long way. The main strain of *Y. pestis* that has been sequenced from aDNA

[12] Cairo was directly affected by the initial strike of the Black Death in the fourteenth century and by many subsequent outbreaks. The modern East African strains, in contrast, all in the 1.ANT group, shared a common ancestor with Central Eurasian strains for, perhaps, a century or more after the Black Death before splitting on a unique evolutionary path.

differs by 45 single-nucleotide polymorphisms (SNPs) from the main Branch 0 lineage in Central Asia (Keller et al. 2019). Although SNPs are not a precise measure of either time or biological circumstances (such as habitat change or a host exchange), they are indicative of the very significant adaptive pressures to which an organism may have had to respond. To give some perspective, the Black Death genome retrieved from western European sites ranging from Barcelona to London and Oslo must have travelled a minimum of several thousand kilometres from the Black Sea, yet it did so with the benefit of the swift maritime technologies of the later Middle Ages that carried it in a matter of months across the Black Sea and the breadth of the Mediterranean. The Black Death genome differs by only two SNPs from the organism that gave rise to the *Y. pestis* polytomy in northern Central Asia sometime in the thirteenth century (Spyrou et al. 2019). Once it arrived in Europe, however, this same strain apparently found new local hosts and had to adapt to them. Genomes from the 1722 plague outbreak in Marseille that descended from the Black Death strain (giving them more than 370 years to develop further) differ by 88 SNPs from their 1348 progenitor. Rather than being simply a function of time or distance, therefore, genetic change in *Y. pestis* lineages seems also to be a function of switching to new intermediary hosts and adapting to new environments.

The 45 SNPs that distinguish the Justinianic Plague genome from the main *Y. pestis* lineage in western China may, therefore, be indicative of host adaptations and climatic alterations that *Y. pestis* endured on its route to Pelusium. So again this raises the question, was that route via the IOW? It has long been recognized that a first-century Greek writer, Rufus of Ephesus, offered a plausible description of bubonic plague infection, suggesting that the disease's presence in or near the Mediterranean Basin may be of considerable antiquity (Green 2014; Mulhall 2019). Rufus placed the disease in Libya, Egypt, and Syria, but those are unlikely to have been long-term reservoirs, since there is no subsequent reference to plague there. The first observations of the Justinianic Plague in Pelusium, rather than in the busier port of Alexandria, implicate the Red Sea as a possible route of disease introduction (Green 2018b). Pelusium was a major receiving and distributing centre for merchandise arriving from Ethiopia, Mesopotamia, and points further east. It was connected by a channel to the Red Sea port of Clysma, which in turn connected the Byzantine Empire and its diplomatic, religious, and trading partners, including the Aksumite Kingdom. The latter was one of the main hubs that linked Indian Ocean trade with the Mediterranean Basin (Selassie 2011; Seland 2014). Epigraphical evidence from South Arabia and contemporary accounts from Byzantine writers pointed to "Ethiopia" as the first noted origin of the plague outbreak, not to the Levant or Persia or the Black Sea as would be more likely for transmission along overland routes, as the conventional narrative suggests (Little 2007, 62–63, 121–122, 249, 304; Green 2014; Harper 2017).

Beaujard, Seland, Campbell, and others have already demonstrated that the ancient world was "globalized" early in the first centuries of the Common Era (Beaujard 2005, 2012; Seland 2008, 2014; Campbell 2016). Long-distance trading systems—aided by the monsoon winds—linked peoples and their diseases in the Roman Empire with those of Africa, Arabia, Persia, India, China, and Central and East Asia by way of the Red Sea, the Persian Gulf, and the Indian Ocean. Moreover, plague can be transmitted across long distances without leaving major human outbreaks in its wake. (In addition to transmission between rodent populations, we need also to consider the possible roles of predators and beasts of burden like camels—which are especially effective transmitters of plague—as well as the human-mediated transport of grain supplies.) It may be quite possible, therefore, to have plague transmission without any trace in the documentary sources. We have, then, a number of possible scenarios in which plague might have been transmitted through the IOW during the First and Second Pandemics, with hints in the genetic evidence of considerable biological phenomena that we cannot yet explain. What is needed now is hard evidence to prove these possibilities one way or the other.

Conclusion

This review of the possible histories of the main infectious diseases of the IOW suggests the following: (1) Malaria—at least vivax malaria—may well have come to the region with hominids migrating from Africa as a "non-specialist" disease, one that could move between various primate hosts, but became exclusively adapted to humans over time. It is likely that prevalence became greater in those areas that farmed rice, but other factors affecting its presence or absence are unclear. The presence of falciparum malaria is not yet documented before the Bronze Age outside of North Africa, whence it extended into the Mediterranean. (2) Tuberculosis seems to have had an African origin, although in this case specifically on the western edge of the IOW. The broad divergence and wide dispersal of the Indian Ocean lineage (Lineage 1) suggests that its movement might have followed trade patterns in the region, but much remains to be investigated. (3) For leprosy, diagnoses from skeletal remains from second or early third millennium BCE India argue for the deep antiquity of that disease in the IOW. The current genetics narrative, however, implicates quite different regions—the Pacific Rim and northern Eurasia—in the earliest history of the disease. The disease does seem to have reached the western rim of the IOW before the end of the first millennium BCE, however, if we can judge from the broad divergence of strains now found in East Africa. (4) If Babkin and Babkina (2012) are correct, smallpox, like tuberculosis, may be a quintessentially IOW disease, originating from the juxtaposition of humans, camels, and naked sole gerbils in the Horn of Africa before spreading around the IOW. (5) Plague, a disease of northern Eurasian origin which was surely the most lethal affliction of pre-modern Afro-Eurasia, may have passed through the IOW at the time of the Justinianic Plague, while in the Second Plague Pandemic, between the fourteenth and eighteenth centuries, it found two separate avenues of entry, via India and (most likely) the Persian Gulf into East Africa. In both regions it focalized, producing strains that persist to the present day.

A dynamic map that traces the movement of the different strains of each of these diseases would likely reveal patterns that mirrored the connectivity that has united this region of the world for centuries. Some diseases remained quite localized in quite small areas. Lineage 7 of the MTBC, for example, has never been found outside of Ethiopia, perhaps because it long ago adapted to hosts with a particular genetic profile. But what patterns of dispersal would be evident if more information were available about the different historical strains of these five diseases? What patterns are there in island versus continental environments, lowlands versus highlands, across paths that followed the monsoons? Indeed, in line with arguments that the monsoons are what make the IOW a unique region in the world, in what ways has the experience of these diseases in the IOW itself been unique? In what ways might slavery systems that relied more on local populations than on long-distance imports have affected disease patterns?[13]

In addition to geography, of course, it is also necessary to identify different chronological strata. For all the sub-regions of the IOW, the fates of

[13] Laso-Jadart et al. (2017) use human genetics to document the importation of a large number of East Africans to what is now Pakistan sometime in the seventeenth or eighteenth century. Although that study does not track any disease transferal, such work may be possible in the near future.

these five diseases changed again in the post-1500 era. The prevalence and, it seems, virulence of smallpox, for example, intensified in all areas of the IOW—and indeed, of the world—in the sixteenth century. The dissemination of European strains of TB altered the dynamics of that disease's effects worldwide. The spread of leprosy in the early modern period is still poorly understood, but by the nineteenth century a global disease crisis was perceived. By any measure, in fact, the nineteenth century was the unhealthiest in world history, with the addition of cholera pandemics to the list of the world's woes. And, of course, just as the new laboratory-based germ theory of disease was developing, the IOW experienced the brunt of the Third Plague Pandemic that originated in Hong Kong in 1894.

New genetics work on the history of infectious diseases is proving transformative for historians. It creates a biological trail that can be analysed alongside traditional historical sources. It provides connections that send historians back to the historical records with new eyes and with new questions. Even though the earliest work in palaeogenetics has been based on samples collected in Europe, the evolutionary approach forces a broader look outside Europe and a closer interrogation of human actions and connections—long-distance migration, trade, and slavery—for the most likely paths of disease dissemination. Future directions in the field of historical epidemiology as it applies to the IOW will necessarily involve bringing molecular genetics more immediately to bear on pathogen isolates, both modern and historical, from this region of the world. Although a survey of all excavated human remains in India, for example, has been published and there is an active cohort of bioarchaeologists working on South Asia (Mushrif-Tripathy et al. 2016; Clark et al. 2017), no aDNA of any of the five pathogenic organisms examined here has yet been retrieved from the IOW. More significantly, too few whole genomes of modern isolates have been sequenced from this part of the world. That absence means that the IOW has thus far been excluded from the developing evolutionary narratives of these pathogens' histories. Having even a dozen whole genome samples of *Y. pestis* from Tanzania and India, where plague remains enzootic (Bertherat 2016), for example, would substantially expand an understanding of how long the disease has been present in those regions.

There are, of course, many infectious diseases that are and have been present in the IOW beyond the five examined here. The most significant omission is cholera, which probably originated in the Ganges Delta. Here, genetics thus far has done less with the history of the pathogen, than with human genetic responses to it. The oldest sequenced sample of *Vibrio cholerae* comes only from the Philadelphia outbreak of 1849 (Devault et al. 2014),[14] and the conventional dating of the cholera pandemics does not begin until 1817. One study suggests, however, the detection of a genetic signal that implies long-term human adaptation to the effects of cholera, allowing populations in Dhaka (Bangladesh) some measure of inherent protection from the disease (Karlsson et al. 2013). Such studies will no doubt continue, and it is necessary that more historians develop skills in interpreting arguments from genetics so that they can better assess the merits or gaps in such studies. At the same time, unsubstantiated theories that link disease origins to the IOW without any method to establish their historicity should be treated with caution. However much the field still has to develop, historical epidemiology now has the IOW on its map. As archaeology, bioarchaeology, genetics, and document-based history forge stronger alliances (Clark et al. 2017), this region of intense historical human activity, migration, and trade—connectivity—will necessarily be incorporated into wider discussions of global health history.

References

Alfani, Guido, and Tommy E. Murphy. 2017. Plague and Lethal Epidemics in the Pre-Industrial World. *The Journal of Economic History* 77 (1): 314–343. https://doi.org/10.1017/S0022050717000092.

Andrades Valtueña, Aida, et al. 2017. The Stone Age Plague and Its Persistence in Eurasia. *Current Biology* 27 (23): 3683–3691.e8. https://doi.org/10.1016/j.cub.2017.10.025.

Arnold, David. 1991. The Indian Ocean as a Disease Zone, 1500–1950. *South Asia: Journal of South Asian Studies* 14 (2): 1–21. https://doi.org/10.1080/00856409108723152.

Babkin, Igor V., and Irina N. Babkina. 2012. A Retrospective Study of the Orthopoxvirus Molecular Evolution. *Infection, Genetics and Evolution* 12 (8): 1597–1604. https://doi.org/10.1016/j.meegid.2012.07.011.

Baker, Brenda J., and Sabrina C. Agarwal. 2017. Stronger Together: Advancing a Global Bioarchaeology. *Bioarchaeology International* 1 (1–2): 1–18. https://doi.org/10.5744/bi.2017.1005.

[14] No other samples of pre-twentieth-century pandemic *V. cholerae* isolates have yet been identified, so no phylogeny extending to the pre-modern era has been attempted.

Barker, Hannah. 2016. Purchasing a Slave in Fourteenth-Century Cairo: Ibn al-Akfānī's Book of Observation and Inspection in the Examination of Slaves. *Mamluk Studies Review* 19: 1–23. https://doi.org/10.6082/M1HH6H5C.

Beaujard, Philippe. 2005. The Indian Ocean in Eurasian and African World-Systems Before the Sixteenth Century. *Journal of World History* 16 (4): 411–465. https://doi.org/10.1353/jwh.2006.0014.

———. 2012. *Les mondes de l'Ocean Indien*. 2 Vols. Paris: Colin.

Benjak, Andrej, et al. 2018. Phylogenomics and Antimicrobial Resistance of the Leprosy Bacillus *Mycobacterium leprae. Nature Communications* 9: 352. https://doi.org/10.1038/s41467-017-02576-z.

Bertherat, E. 2016. Plague Around the World, 2010–2015. *Weekly Epidemiological Record/Relevé épidémiologique hebdomadaire* 91 (8): 89–104.

Bos, Kirsten I., et al. 2014. Pre-Columbian Mycobacterial Genomes Reveal Seals as a Source of New World Human Tuberculosis. *Nature* 514: 494–497. https://doi.org/10.1038/nature13591.

Buckley, Hallie R. 2006. 'The Predators Within': Investigating the Relationship Between Malaria and Health in the Prehistoric Pacific Islands. In *Bioarchaeology of Southeast Asia*, ed. M. Oxenham and N. Tayles, 309–332. Cambridge: Cambridge University Press.

Cambridge University Library. 2017. Medicine in Medieval Egypt. http://www.lib.cam.ac.uk/collections/departments/taylor-schechter-genizah-research-unit/projects/medicine-medieval-egypt. Accessed 6 July 2017

Campbell, Gwyn. 2011. Slavery in the Indian Ocean World. In *The Routledge History of Slavery*, ed. Gad Heuman and Trevor Burnard, 52–63. London: Routledge.

———, ed. 2016. *Early Exchange Between Africa and the Wider Indian Ocean World*. London: Palgrave Macmillan.

Casale, Giancarlo. 2010. *The Ottoman Age of Exploration*. Oxford: Oxford University Press.

Chevers, Norman. 1886. *A Commentary on the Diseases of India*. London: J. & A. Churchill.

Clark, Angela L., et al. 2017. Biological Anthropology in the Indo-Pacific Region: New Approaches to Age-Old Questions. *Journal of Indo – Pacific Archaeology* 41: 78–94. https://doi.org/10.7152/jipa.v41i0.15021.

Comas, Iñaki, et al. 2015. Population Genomics of *Mycobacterium tuberculosis* in Ethiopia Contradicts the Virgin Soil Hypothesis for Human Tuberculosis in Sub-Saharan Africa. *Current Biology* 25 (24): 3260–3266. https://doi.org/10.1016/j.cub.2015.10.061.

Coscolla, Mireia, and Sebastien Gagneux. 2014. Consequences of Genomic Diversity in *Mycobacterium tuberculosis. Seminars in Immunology* 26: 431–444. https://doi.org/10.1016/j.smim.2014.09.012.

Crosby, Alfred W. 1972. *The Columbian Exchange: Biological and Cultural Consequences of 1492*. Westport, CT: Greenwood.

———. 1986. *Ecological Imperialism: The Biological Expansion of Europe, 900–1900*. Cambridge: Cambridge University Press.

Cui, Yujun, et al. 2013. Historical Variations in Mutation Rate in an Epidemic Pathogen, *Yersinia pestis*. *Proceedings of the National Academy of Science* 110 (2): 577–582. https://doi.org/10.1073/pnas.1205750110.

Curtin, Philip D. 1961. 'The White Man's Grave': Image and Reality, 1780–1850. *The Journal of British Studies* 1 (1): 94–110. https://doi.org/10.1086/385437.

———. 1968. Epidemiology and the Slave Trade. *Political Science Quarterly* 83 (2): 190–216. https://doi.org/10.2307/2147089.

Derat, Marie-Laure. 2018. Du lexique aux talismans: occurrences de la peste dans la Corne de l'Afrique du XIIIe au XVe siècle. *Afriques* 9. https://doi.org/10.4000/afriques.2090.

Devault, Alison M., et al. 2014. Second-Pandemic Strain of *Vibrio cholerae* from the Philadelphia Cholera Outbreak of 1849. *New England Journal of Medicine* 370: 334–340. https://doi.org/10.1056/NEJMoa1308663.

Dols, Michael W. 1977. *The Black Death in the Middle East*. Princeton: Princeton University Press.

Duggan, Ana T., et al. 2016. 17th Century Variola Virus Reveals the Recent History of Smallpox. *Current Biology* 26: 3407–3412. https://doi.org/10.1016/j.cub.2016.10.061.

Emmerick, Roland E. 1984. Some Remarks on the History of Leprosy in India. *Indologica Taurinensia* 12: 93–105.

Gagneux, Sebastien. 2012. Host-pathogen Coevolution in Human Tuberculosis. *Philosophical Transactions of the Royal Society. B. Biological Sciences* 367: 850–859. https://doi.org/10.1098/rstb.2011.0316.

———. 2018. Ecology and Evolution of *Mycobacterium tuberculosis*. *Nature Reviews*, published online 19 Feb 2018. https://doi.org/10.1038/nrmicro.2018.8

Gething, Peter W., et al. 2011. A New World Malaria Map: *Plasmodium falciparum* endemicity in 2010. *Malaria Journal* 10: 378. https://doi.org/10.1186/1475-2875-10-378.

———. 2012. A Long Neglected World Malaria Map: *Plasmodium vivax* Endemicity in 2010. *PLoS Neglected Tropical Diseases* 6 (9): e1814. https://doi.org/10.1371/journal.pntd.0001814.

Goitein, S.D., and M. Friedman, ed. and trans. 2008. *India Traders of the Middle Ages: Documents from the Cairo Geniza ('India Book')*. Leiden: Brill.

Green, Monica H. 2014. Taking 'Pandemic' Seriously: Making the Black Death Global. *The Medieval Globe* 1 (1): 27–61. http://scholarworks.wmich.edu/tmg/vol1/iss1/4.

———. 2017. The Globalisations of Disease. In *Human Dispersal and Species Movement: From Prehistory to the Present*, ed. Nicole Boivin, Rémy Crassard, and Michael D. Petraglia, 494–520. Cambridge: Cambridge University Press. https://doi.org/10.1017/9781316686942.020.

———. 2018a. Climate and Disease in Medieval Eurasia. In *Oxford Research Encyclopedia of Asian History*. Oxford: Oxford University Press. https://doi.org/10.1093/acrefore/9780190277727.013.6.

———. 2018b. Putting Africa on the Black Death Map: Narratives from Genetics and History. *Afriques* 9. http://journals.openedition.org/afriques/.

Gupta, Manohar Lal, and Anuradha Sharma. 2007. Pneumonic Plague, Northern India, 2002. *Emerging Infectious Diseases* 13 (4): 664–666. https://doi.org/10.3201/eid1304.051105.

Han, Xiang Y. 2017. Detection of the Leprosy Agent *Mycobacterium lepromatosis* in South America and Europe. *American Journal of Tropical Medicine and Hygiene* 96 (1): 260. https://doi.org/10.4269/ajtmh.16-0713.

Harper, Kyle. 2017. *The Fate of Rome: Climate, Disease, and the End of an Empire*. Princeton: Princeton University Press.

Harrison, Mark. 2015. A Global Perspective: Reframing the History of Health, Medicine, and Disease. *Bulletin of the History of Medicine* 89 (4): 639–689. https://doi.org/10.1353/bhm.2015.0116.

Harrison, Mark, and Michael Worboys. 1997. A Disease of Civilisation: Tuberculosis in Britain, Africa and India, 1900–39. In *Migrants, Minorities and Health: Historical and Contemporary Studies*, ed. Lara Marks and Michael Worboys, 93–124. New York: Routledge.

Honap, Tanvi, et al. 2017. Genomic Analyses of Ancient *Mycobacterium tuberculosis* Complex Strains from the Americas. Posters presented at the 2017 Max Planck Institut für Evolutionsbiologie meeting on "One Past Health" (http://pubman.mpdl.mpg.de/pubman/faces/viewItemOverviewPage.jsp?itemId=escidoc:2426270), and the 2017 meeting of the Society for Molecular Biology and Evolution (http://www.smbe2017.org/wp-content/uploads/2016/02/SMBE-Abstract-Book-22-June.pdf).

Hoogervorst, Tom. 2016. Tracing Maritime Connections Between Island Southeast Asia and the Indian Ocean World. In *The Routledge Handbook of Archaeology and Globalization*, ed. Tamar Hodos, 751–767. New York: Routledge. https://doi.org/10.4324/9781315449005.ch51.

Horton, Mark. 2017. Early Islam on the East African Coast. In *A Companion to Islamic Art and Architecture*, ed. Finbarr Barry Flood and Gulru Necipoglu, 250–274. Hoboken, NJ: John Wiley and Sons.

Houben, Rein M.G.J., and Peter J. Dodd. 2016. The Global Burden of Latent Tuberculosis Infection: A Re-estimation Using Mathematical Modelling. *PLoS Medicine* 13 (10): e1002152. https://doi.org/10.1371/journal.pmed.1002152.

Houldcroft, Charlotte J., and Simon J. Underdown. 2016. Neanderthal Genomics Suggests a Pleistocene Time Frame for the First Epidemiologic Transition. *American Journal of Physical Anthropology* 160: 379–388. https://doi.org/10.1002/ajpa.22985.

Hymes, Robert. 2014. A Hypothesis on the East Asian Beginnings of the *Yersinia pestis* Polytomy. *The Medieval Globe* 1: 285–308.

Jahangir. 1909–1914. *The Tuzuk-i-Jahangiri; Or, Memoirs of Jahangir*. Trans. Alexander Rogers and Ed. Henry Beveridge. London: London Royal Asiatic Society.

Karlsson, Elinor K., et al. 2013. Natural Selection in a Bangladeshi Population from the Cholera-Endemic Ganges River Delta. *Science Translational Medicine* 5 (192): 192ra86. https://doi.org/10.1126/scitranslmed.3006338.

Kay, Gemma L., et al. 2015. Eighteenth-Century Genomes Show That Mixed Infections Were Common at Time of Peak Tuberculosis in Europe. *Nature Communications* 6: 6717. https://doi.org/10.1038/ncomms7717.

Keller, Marcel, et al. 2019. Ancient *Yersinia pestis* Genomes from Across Western Europe Reveal Early Diversification During the First Pandemic (541–750). *Proceedings of the National Academy of Sciences* 116 (25): 12363–12372. https://doi.org/10.1073/pnas.1820447116.

King, Charlotte L., Siân E. Halcrow, Nancy Tayles, and Stephanie Shkrum. 2017. Considering the Palaeoepidemiological Implications of Socioeconomic and Environmental Change in Southeast Asia. *Archaeological Research in Asia* 11: 27–37. https://doi.org/10.1016/j.ara.2017.05.003.

Kingston, Joseph J., et al. 2009. Genotyping of Indian *Yersinia pestis* Strains by MLVA and Repetitive DNA Sequence Based PCRs. *Antonie van Leeuwenhoek* 96: 303–312. https://doi.org/10.1007/s10482-009-9347-2.

Larremore, Daniel B., et al. 2015. Ape Parasite Origins of Human Malaria Virulence Genes. *Nature Communications* 6: 8368. https://doi.org/10.1038/ncomms9368.

Laso-Jadart, Romuald, et al. 2017. The Genetic Legacy of the Indian Ocean Slave Trade: Recent Admixture and Post-admixture Selection in the Makranis of Pakistan. *American Journal of Human Genetics* 101 (6): 977–984. https://doi.org/10.1016/j.ajhg.2017.09.025.

Leung, Angela Ki Che. 2009. *Leprosy in China: A History*. New York: Columbia University Press.

Little, Lester K., ed. 2007. *Plague and the End of Antiquity: The Pandemic of 541–750*. Cambridge: Cambridge University Press.

Liu, Weimin, et al. 2010. Origin of the Human Malaria Parasite *Plasmodium falciparum* in Gorillas. *Nature* 467 (7314): 420–425. https://doi.org/10.1038/nature09442.

———. 2014. African Origin of the Malaria Parasite *Plasmodium vivax*. *Nature Communications* 5: 3346. https://doi.org/10.1038/ncomms4346.

Loy, Dorothy E., et al. 2017. Out of Africa: Origins and Evolution of the Human Malaria Parasites *Plasmodium falciparum* and *Plasmodium vivax*. *International Journal of Parasitology* 47: 87–97. https://doi.org/10.1016/j.ijpara.2016.05.008.

Luo, Tao, et al. 2014. Southern East Asian Origin and Coexpansion of *Mycobacterium tuberculosis* Beijing Family with Han Chinese. *PNAS* 112 (26): 8136–8141. https://doi.org/10.1073/pnas.1424063112.

Mahale, Kiran N., et al. 2014. Draft Genome Sequences of *Yersinia pestis* Strains from the 1994 Plague Epidemic of Surat and 2002 Shimla Outbreak in India. *Indian Journal of Microbiology* 54 (4): 480–482. https://doi.org/10.1007/s12088-014-0475-7.

Marciniak, Stephanie, et al. 2018. A Multi-Faceted Anthropological and Genomic Approach to Framing *Plasmodium falciparum* Malaria in Imperial Period Central-Southern Italy (1st–4th c. CE). *Journal of Anthropological Archaeology* 49: 210–224. https://doi.org/10.1016/j.jaa.2018.01.004.

Mark, Samuel. 2002. Alexander the Great, Seafaring, and the Spread of Leprosy. *Journal of the History of Medicine and Allied Sciences* 57 (3): 285–311. muse.jhu.edu/article/15345.

McNeill, William H. 1976. *Plagues and Peoples.* Garden City, NY: Anchor Press.

McNeill, John R. 2010. *Mosquito Empires: Ecology and War in the Greater Caribbean, 1640–1914.* Cambridge and New York: Cambridge University Press.

Meulenbeld, G. (1999–2002, January). *A History of Indian Medical Literature.* 5 Vols. Groningen: Egbert Forsten.

Meyer, Matthias, et al. 2016. Nuclear DNA Sequences from the Middle Pleistocene Sima de los Huesos Hominins. *Nature* 531 (7595): 504–507. https://doi.org/10.1038/nature17405.

Mitchell, Stephen. 2014. *A History of the Later Roman Empire, AD 284–641.* 2nd ed. New York: Wiley-Blackwell.

Morelli, Giovanna, et al. 2010. *Yersinia pestis* Genome Sequencing Identifies Patterns of Global Phylogenetic Diversity. *Nature Genetics* 42 (12): 1140–1145. Supplementary Table 4. https://doi.org/10.1038/ng.705.

Mulhall, John. 2019. Plague Before the Pandemics: The Greek Medical Evidence for Bubonic Plague Before the Sixth Century. *Bulletin of the History of Medicine* 93 (2): 151–179. https://doi.org/10.1353/bhm.2019.0025.

Mushrif-Tripathy, Veena, K.S. Chakraborty, and S. Lahiri. 2016. Where Are They Now? The Human Skeletal Remains from India. In *A Companion to South Asia in the Past*, ed. Gwen Robbins Schug and Subhash R. Walimbe, 496–533. Hoboken, NJ: John Wiley and Sons. https://doi.org/10.1002/9781119055280.ch32.

Newfield, Timothy P. 2017. Malaria and Malaria-Like Disease in the Early Middle Ages. *Early Medieval Europe* 25 (3): 251–300. https://doi.org/10.1111/emed.12212.

O'Neill, Mary B., et al. 2017. Lineage Specific Histories of *Mycobacterium tuberculosis* Dispersal in Africa and Eurasia. *bioRxiv*, first posted online October 27, 2017. https://doi.org/10.1101/210161.

Rasmussen, Simon, et al. 2015. Early Divergent Strains of *Yersinia Pestis* in Eurasia 5,000 Years Ago. *Cell* 163 (3): 571–582. https://doi.org/10.1016/j.cell.2015.10.009.

Robbins Schug, Gwen. 2016. Begotten of Corruption? Bioarchaeology and 'Othering' of Leprosy in South Asia. *International Journal of Paleopathology* 15: 1–9. https://doi.org/10.1016/j.ijpp.2016.09.002.

Robbins Schug, Gwen, and Kelly Elaine Blevins. 2016. The Center Cannot Hold: A Bioarchaeological Perspective on Environmental Crisis in the Second Millennium BCE, South Asia. In *A Companion to South Asia in the Past*, ed. Gwen Robbins Schug and Subhash R. Walimbe, 255–273. Hoboken, NJ: John Wiley and Sons. https://doi.org/10.1002/9781119055280.ch16.

Robbins Schug, Gwen, et al. 2013. Infection, Disease, and Biosocial Processes at the End of the Indus Civilization. *PLoS ONE* 8 (12): e84814. https://doi.org/10.1371/journal.pone.0084814.

Robbins, Gwen, et al. 2009. Ancient Skeletal Evidence for Leprosy in India (2000 B.C.). *PLoS One* 4 (5): 1–8. https://doi.org/10.1371/journal.pone.0005669.

Roberts, Charlotte A., and Jane E. Buikstra, eds. 2003. *The Bioarchaeology of Tuberculosis: A Global View on a Re-emerging Disease*. Gainesville: University Press of Florida.

Roser, Max. 2016. Smallpox Was Globally Eradicated in 1977 – This Map Shows the Year of Eradication of Smallpox. *Our World in Data*. https://ourworldindata.org/roser/maps/YearOfSmallpoxEradication/YearOfSmallpoxEradication.html. Accessed 12 July 2017.

Royer, Katherine. 2014. The Blind Men and the Elephant: Imperial Medicine, Medieval Historians, and the Role of Rats in the Historiography of Plague. In *Medicine and Colonialism: Historical Perspectives in India and South Africa*, ed. Poonam Bala, 99–110. London: Pickering & Chatto.

Schuenemann, Verena J., et al. 2013. Genome-wide Comparison of Medieval and Modern *Mycobacterium leprae*. *Science* 341: 179–183. https://doi.org/10.1126/science.1238286.

———. 2018. Ancient Genomes Reveal a High Diversity of *Mycobacterium leprae* in Medieval Europe. *PLOS Pathogens* 14 (5): e1006997. https://doi.org/10.1371/journal.ppat.1006997.

Seland, Eivind Heldaas. 2008. The Indian Ocean and the Globalisation of the Ancient World. *Ancient West and East* 7: 67–79. https://doi.org/10.2143/AWE.7.0.2033253.

———. 2014. Archaeology of Trade in the Western Indian Ocean, 300 BC–AD 700. *Journal of Archaeological Research* 22: 367–402. https://doi.org/10.1007/s10814-014-9075-7.

Selassie, Yohannes Gebre. 2011. Plague as a Possible Factor for the Decline and Collapse of the Aksumite Empire: A New Interpretation. *ITYOPIS—Northeast African Journal of Social Sciences and Humanities* 1: 36–61. http://www.ityopis.org/Issues-1_files/ITYOPIS-I-Gebre-Selassi.pdf.

Sharif, M. 1951. Spread of Plague in the Southern and Central Divisions of Bombay Province and Plague Endemic Centres in the Indo-Pakistan Subcontinent. *Bulletin of the World Health Organization* 4: 75–109.

Singh, Pushpendra, et al. 2015. Insight into the Evolution and Origin of Leprosy Bacilli from the Genome Sequence of *Mycobacterium lepromatosis*. *PNAS* 112 (14): 4459–4464. https://doi.org/10.1073/pnas.1421504112.

Smith, Cairns S., et al. 2017. Multidrug Therapy for Leprosy: A Game Changer on the Path to Elimination. *The Lancet Infectious Diseases* 17 (9): e293–e297. https://doi.org/10.1016/S1473-3099(17)30418-8.

Sparacello, Vitale Stefano, et al. 2016. Insights on the Paleoepidemiology of Ancient Tuberculosis from the Structural Analysis of Postcranial Remains from the Ligurian Neolithic (Northwestern Italy). *International Journal of Paleopathology* 15: 50–64. https://doi.org/10.1016/j.ijpp.2016.08.003.

Spyrou, Maria A., et al. 2016. Historical *Y. pestis* Genomes Reveal the European Black Death as the Source of Ancient and Modern Plague Pandemics. *Cell Host and Microbe* 19 (6): 874–881. https://doi.org/10.1016/j.chom.2016.05.012.

Spyrou, Maria A., et al. 2019. Phylogeography of the Second Plague Pandemic Revealed Through Analysis of Historical *Yersinia Pestis* Genomes. *Nature Communications* 10 (1): 4470. https://doi.org/10.1038/s41467-019-12154-0.

Stucki, David, et al. 2016. *Mycobacterium tuberculosis* Lineage 4 Comprises Globally Distributed and Geographically Restricted Sublineages. *Nature Genetics* 48 (12): 1535–1543. https://doi.org/10.1038/ng.3704.

Sundararaman, Sesh A., et al. 2016. Genomes of Cryptic Chimpanzee *Plasmodium* Species Reveal Key Evolutionary Events Leading to Human Malaria. *Nature Communications* 7: 11078. https://doi.org/10.1038/ncomms11078.

Supply, Philip, and Roland Brosch. 2017. The Biology and Epidemiology of *Mycobacterium canettii*. In *Strain Variation in the Mycobacterium tuberculosis Complex: Its Role in Biology, Epidemiology and Control*, ed. Sebastien Gagneux, 27–41. Berlin/New York: Springer. https://doi.org/10.1007/978-3-319-64371-7_2.

Sussman, George D. 2011. Was the Black Death in India and China? *Bulletin of the History of Medicine* 85 (3): 319–355. https://doi.org/10.1353/bhm.2011.0054.

———. 2016. Scientists Doing History: Central Africa and the Origins of the First Plague Pandemic. *Journal of World History* 26: 325–354. https://doi.org/10.1353/jwh.2016.0030.

Tayles, N., and H.R. Buckley. 2004. Leprosy and Tuberculosis in Iron Age Southeast Asia? *American Journal of Physical Anthropology* 125: 239–256. https://doi.org/10.1002/ajpa.10378.

Thin, George. 1891. Map Showing the Distribution of Leprosy around the World in 1891. In *Leprosy*, ed. George Thin. London: Percival and Co. https://wellcomeimages.org/image#L0032805. Accessed 19 July 2017.

Van Linschoten, Jan Huyghen. 1598/1885. In *The Voyage of John Huyghen van Linschoten to the East Indies: From the Old English Translation of 1598: The First Book, Containing His Description of the East*, ed. Arthur Coke Burnell and P.A. Tiele. London: Hakluyt Society.

Varlık, Nükhet. 2015. *Plague and Empire in the Early Modern Mediterranean World: The Ottoman Experience, 1347–1600*. Cambridge: Cambridge University Press.

Viganó, Claudia, et al. 2017. 2,000 Year Old Β-Thalassemia Case in Sardinia Suggests Malaria Was Endemic by the Roman Period. *American Journal of Physical Anthropology* 164: 362–370. https://doi.org/10.1002/ajpa.23278.

Wagner, David M., et al. 2014. *Yersinia pestis* and the Plague of Justinian 541–543 AD: A Genomic Analysis. *The Lancet Infectious Diseases* 14 (4): 319–336. https://doi.org/10.1016/S1473-3099(13)70323-2.

Webb, James L.A., Jr. 2009. *Humanity's Burden: A Global History of Malaria*. New York: Cambridge University Press.

———. 2017. Early Malarial Infections and the First Epidemiological Transition. In *Human Dispersal and Species Movement: From Prehistory to the Present*, ed. Nicole Boivin, Rémy Crassard, and Michael D. Petraglia, 477–493. Cambridge: Cambridge University Press.

World Health Organization. 2008. *Interregional Meeting on Prevention and Control of Plague, Antananarivo, Madagascar, 1–11 April 2006*. Geneva: WHO.

Wujastyk, Dominik, Anthony Cerulli, and Karin Preisendanz, eds. 2013. *Medical Texts and Manuscripts in Indian Cultural History*. New Delhi: Manohar.

Wynne-Jones, Stephanie, and Adria LaViolette. 2017. *The Swahili World*. New York: Routledge.

Zehender, Gianguglielmo, et al. 2018. Bayesian Reconstruction of the Evolutionary History and Cross-Species Transition of Variola Virus and Orthopoxviruses. *Journal of Medical Virology*, accepted manuscript online: 15 February 2018. https://doi.org/10.1002/jmv.25055.

Ziegler, Michelle. 2016. Malarial Landscapes in the Late Antique Tiber Valley. *Landscapes* 17 (2): 139–155. https://doi.org/10.1080/14662035.2016.1251041.

CHAPTER 3

The 'Frankish Disease' and Its Treatments in the Indian Ocean World

Anna Winterbottom

Introduction

Towards the end of the fifteenth century, people in many different parts of the world began to complain of sores, ulcers and rashes, sometimes followed by decay of the bones and even madness. These symptoms were allegedly caused by close contacts with foreigners. Across much of the Indian Ocean world (IOW), they were referred to using various derivatives of *afrang*, an old Persian term for 'Frank', including *farangi* and *firingi*. A similar term, *parangi*, used in parts of India and Sri Lanka to refer to the Portuguese, was also used to designate the new disease. Accordingly, I have used the term 'Frankish disease' here as a general translation of these various terms.

The Frankish disease is usually identified with the contemporary 'Great Pox' in Europe,[1] which in turn is often equated with the modern disease

[1] For the term 'Great Pox' and other names for venereal disease in early modern Europe, see Arrizabalaga et al. (1997).

A. Winterbottom (✉)
McGill University, Montreal, QC, Canada

G. Campbell, E.-M. Knoll (eds.), *Disease Dispersion and Impact in the Indian Ocean World*, Palgrave Series in Indian Ocean World Studies, https://doi.org/10.1007/978-3-030-36264-5_3

syphilis. However, retrospective diagnosis of conditions reported before germ theory presents several problems (Arrizabalaga 2002; Cooter and Stein 2013). Diseases caused by the bacterium *Treponema pallidum* are known as treponemal diseases. The modern term 'syphilis' refers to the disease caused by a particular strain of this bacterium. In contrast, the historical terms mentioned above described a variable collection of symptoms. Several could have had a number of causes, including other strains of *T. pallidum*, like those that cause the non-venereal conditions yaws, pinta and bejel. Before the twentieth century, syphilis was not clearly differentiated from gonorrhoea, though the latter is caused by a different bacterium, *Neisseria gonorrhoeae*. The Frankish disease also reportedly resembled other conditions, notably the so-called Persian fire, which it is difficult to identify clearly with any modern disease.

Given the problems with retrospective diagnoses, it might be questioned whether it is justifiable to discuss historical epidemics at all. However, the borrowing of terminology across cultures does suggest a contemporary understanding of a single phenomenon, even if the symptoms that were considered to make up the disease or the experience of illness varied from place to place. In this chapter, therefore, I shall take the Frankish disease on its own terms, noting how it was described by contemporaries, including its supposed causes, the humoral profiles that were assigned to it in different societies around the IOW and the ways in which it was understood to be communicated. I shall also examine how understandings of disease interacted with ideas about society, particularly views of foreigners and others, above all prostitutes, who were tainted with the suspicion of contagion. Turning then to treatments for the new malady, I shall ask why some remained local while others spread quickly across the globe, being adopted by far-flung societies with apparently quite different understandings of the body and healing. Finally, I shall briefly discuss venereal disease as imagined by the European colonists in the IOW from around 1800 to 1900.

The overall aim of this study of the 'Frankish disease' is to answer a basic question about the integrity of the Indian Ocean region as a unit of historical analysis, or a 'world', as I and others in this volume and elsewhere have referred to it. Can we move beyond recognizing the region as a 'disease zone' (Arnold 1991) and demonstrate commonalities and connectivity in how people understood disease and practised healing?

Historiography

There is a long-running debate over whether the Frankish disease was new to the early modern world. Approaches have included examining historical records, skeletal records and most recently DNA studies. The 'Columbian exchange' theory, first suggested in the sixteenth century and more recently championed by Alfred Crosby (1969, 1972), is that syphilis was an American disease that arrived in the Old World with Columbus' returning fleet in October 1492 (Harper et al. 2011). Others argue that syphilis and other treponemal diseases had long been known across the world but that they were misdiagnosed as other conditions, including leprosy.

Alternative theories include the 'unitarian' theory, which originally argued that all treponemal diseases have a single cause and that syphilis evolved from a strain passed on through non-sexual bodily contact, perhaps in response to changing environmental conditions or behavioural changes in humans (Powell and Cook 2005).[2] During the 1980s, genetic studies added support to this argument by establishing that syphilis is caused by just one form of *T. pallidum*. Recent whole-genome sequencing studies have confirmed that syphilis and yaws bear a close genetic resemblance (Harper et al. 2014).

Some recent genetic studies have suggested a consensus: the bacterium that caused yaws mutated into that which causes syphilis, but this probably happened first in the Americas, the new bacterium then being transmitted to the Old World in the late fifteenth century (ibid.). This theory remains contested, however, by those who regard the different strains of the *T. pallidum* bacterium as having emerged at around the same time.

The result of this long debate is that most historical studies of the 'Great Pox' in the early modern period have concentrated on trying to prove or disprove the Columbian exchange theory. More recent historical scholarship on the 'Great Pox' in Europe has moved away from this perspective towards a socio-historical approach, asking what the disease revealed about society and how it changed medicine itself (Arrizabalaga et al. 1997). In contrast, there are few detailed accounts of the contempo-

[2] The non-venereal treponemal diseases are passed on by contact with fluid from the lesion of an infected person. Behavioural changes that could limit the opportunities for this mode of transmission include better sanitation or more clothing. Increased sexual contact has also been suggested as a behavioural change that might have encouraged the venereal transmission of disease. The non-venereal diseases are usually present in tropical climates, so the bacteria could have adapted to sexual transmission in order to survive in colder environments.

rary epidemic in Africa, Asia or the Middle East. A notable exception is Keizō Dohi's work, originally published in Japanese in 1921 and translated into German shortly afterwards (Dohi 1923). Dohi covers Asia in some detail and shows how the points of infection around the Indian Ocean can be mapped on to the trajectories of European voyages of discovery. However, he does not cover Africa or the Middle East in any detail.

Since Dohi's work, regional and local studies have uncovered many more references to the Frankish disease in the IOW. In addition, detailed accounts have been given of late colonial efforts to eradicate venereal disease in the region. While scholarship on the early modern period generally focuses on understanding the dates and geographical dimensions of the epidemic, historians of the later period are generally more attuned to the social and racial aspects of ideas about disease and contagion and tend to pay less attention to the question of the actual extent of disease. Here, I will draw together some of these works with the aim of integrating both the information they provide and their theoretical perspectives.

Sources

Sources for the Frankish disease in the IOW include works by medical authors, chroniclers, poets, historians and autobiographers. European travellers to the region also described the disease.

The writer known as Leo Africanus (al-Ḥasan ibn Muḥammad al-Wazzān al-Zayyātī or al-Fāsī, c. 1485–1554) was born in Granada, in Al-Andalus or Muslim Spain, but spent most of his life in North Africa before being captured by Christian pirates and presented to Pope Leo X. In around 1526, while living in Rome, he composed his *History of Africa* (Épaulard 1960, 60), in which he describes the *mal français* (French disease) spreading throughout much of North Africa. He attributed its spread to the expulsion of the Jews from Spain and in particular to Arab men having sexual intercourse with Jewish women.[3] This account is problematic from the viewpoint of the Columbian exchange theory, because the charter issued by Ferdinand and Isabella commanding the expulsion of all unconverted Jews was issued on 31 March 1492, and the exodus was largely complete by July the same year, shortly before Columbus' return. However, the account is typical in that it ascribes the

[3] According to Arrizabalaga et al. (1997, 14), in 1789 Christian Gottfried Gruner argued that the disease originated among both the Arabs and the Jews expelled from Spain in 1492.

Frankish disease to a period of conflict and forced migration and to the influx of foreigners, foreign women being considered particularly dangerous.

In Egypt, an epidemic of *ḥabb ifranji* (Frankish pustule) was mentioned by Muḥammad ibn Aḥmad Ibn Iyās (c. 1447–1524), a writer of Mamluk descent living in Cairo. In the first entry in his chronicle, Ibn Iyās dated the arrival of the disease to 1495. He describes its severity and the inability of doctors to cure or alleviate it. Dāʾūd al-Anṭākī (died 1599), a physician from Antioch who worked at Cairo hospital in the late sixteenth century, was the first to describe the epidemic and its treatments in detail (Bachour 2015; Renaud and Colin 1935). Al-Anṭākī dated the appearance of the Frankish pustule to 807 AH (1404–1405 CE), although most commentators have agreed that this must be an error for 907 AH (1501–1502 CE). At around the same time, a little-known Moroccan poet, ʾAbd al-Karim b. Muʾmin b. Yahya (active c. 1557–1574), composed a poem on a disease he called *bibas* (after the Spanish word *bubas*, meaning literally 'pustules'), in which he mentioned treatments for it using arsenic and mercury (Renaud and Colin 1935).

The Frankish disease might have spread into Arabia either from Egypt under the Mamluk Sultanate or as a result of the Portuguese presence on the South Arabian coast. According to Ibn Iyās, several troops appealed to be excused from a Mamluk expedition to capture Yemen as part of a planned counter-attack on Portuguese forces, on account of their having the 'Frankish chancre'. Chronicle entries from the Hadhramaut dated 906 AH (1500–1501 CE) or 909 AH (1503–1504 CE) mention a bad sort of 'Persian fire' (Serjeant 1965). Persian fire was a condition described from the time of the famed philosopher-scientist Abū ʿAlī al-Ḥusayn ibn ʿAbd Allāh ibn Sīnā (usually known as Ibn Sīnā or in Latin, Avicenna, 980–1037 CE) onwards. It apparently caused swellings of the skin (pustules or vesicular eruptions) which some have identified with carbuncles or anthrax (Elgood 1951, 376). However, since many different conditions can cause such symptoms, it is difficult to identify Persian fire with any modern condition.

In Persia, the Frankish disease was first described by Bahāʾ ad-Dawla Rāzī (died 1507), a medical writer from the town of Rayy who had studied with Indian medical scholars in Herat, now in Afghanistan. In his *Ḫulāṣatat-taǧārib* ('Essence of the Experiments') of 1501, Bahāʾ ad-Dawla said that the new disease was known variously as *ābila-i Farang* (Frankish smallpox), *armānī-I dāna* (Armenian grain) or *ātašak* (little fire)

(Thomann 2015). The last name, which became standard in Persian, was derived from a comparison with Persian fire, which, like the new disease, caused a burning sensation (ibid.). The early sixteenth-century poet and physician, Yūsuf Ibn Muḥammad Ibn Yūsuf (called Yūsufi, active c. 1507–1539), described the 'Frankish smallpox' and its treatments in his *Gami' al-fawa'id* (c. 1511), as did ʿImād ad-Dīn Šīrāzī in his *Risāla-i ātašak*, or 'Epistle of the little fire', composed in 1569 (Elgood 1931, 1951; Thomann 2015). ʿImād ad-Dīn, whose own dates of birth and death are unknown, belonged to a prominent family of physicians and himself served the court of the Safavid ruler Shah Ṭahmāsp (1524–1576). He later lived in Mashhad in eastern Iran, an important centre for pilgrimage at the time, where he worked at the shrine hospital of Imam Reza (Thomann 2015; Savage-Smith 1998). Sites of pilgrimage were often centres for prostitution and hence often had high rates of venereal disease, and Mashhad was known for its legalized prostitution (Thomann 2015). ʿImād ad-Dīn did not mention this himself, but it might explain his interest in the 'little fire'.

ʿImād ad-Dīn is categorical in viewing the 'little fire' as a new phenomenon, arguing that it arrived in Persia via Armenia. In early modern Islamic medicine, the concept of contagion was contentious, following a tradition that the Prophet Muhammad had dismissed it as a superstition, teaching instead that all disease comes from God (Conrad 1992). In practice, however, many medical writers accepted the possibility that diseases could be transmitted from one person to another (Stearns 2011). Thus, ʿImād ad-Dīn categorized the disease as arising from black bile, but he also cautioned that it could be infectious, notably through sexual contact. He also mentioned sharing blood and hot baths as means of transmission.

This partial acceptance of the concept of contagion alongside humoral explanation in the Islamic world was similar to the situation in Europe. Epidemic disease, notably plague, had been one factor in encouraging the accommodation between popular ideas of contagion and humoral medicine. It might be significant that the medical writer who first proposed a theory of 'seeds of disease' or germs, Girolamo Fracastoro, was also the coiner of the term 'syphilis' (Nutton 1990). However, various theories of contagion had co-existed with humoral and environmental explanations for diseases for much longer in both European and Islamic medical cultures, also having precedents in the writings of ancient authors, including Galen (ibid.; Stearns 2011).

Like many of his European contemporaries, ʿImād ad-Dīn associated the new disease primarily with black bile (also the humour thought to be responsible for madness) that, he argued, had become infected or 'fermented' (Elgood 1931). Black bile was often said to be a sedimentary black substance produced in the process of the manufacture of blood, some of which remained in the blood and some in the spleen. Like the other humours, black bile was thought to undergo physical processes such as becoming putrid or burnt, with various effects on the body and mind (Savage-Smith 2013).

Other Persian treatises on the 'little fire' followed, but that of ʿImād ad-Dīn remained influential, forming the basis for several later works in both Persian and Arabic. Just as in North Africa, a few Persian accounts place the arrival of the disease earlier than the 1490s. In 1514–1515, for example, Sultān 'Alī, then about 74 years old, complained of having suffered 'the grievous Frankish scourge' in his youth.

The earliest evidence of the appearance of the Frankish disease in South Asia probably comes from the Italian traveller Ludovico di Varthema (c. 1465–1517). Writing in 1505 (Varthema 1928), he reported that the 'Great Pox' had arrived in India with the early Portuguese colonists by the end of the fifteenth century. Varthema reported that the disease affected the ruler of Calicut in his throat (a common place for an early chancre or lesion to appear in cases of syphilis).

The account given by Bhāvamiśra (active c. 1550–1590) in his Sanskrit medical treatise *Bhāvaprakāśa*, normally dated to 1558, is probably the earliest description of the new disease to appear in Indian medical literature. Bhāvamiśra referred to it as *phiraṅga roga* (the disease of the Franks), and he specified that it was contracted from sexual contact with foreigners from the West, where he believed the disease to have originated (Bhāvamiśra 1998; Wujastyk 2003, 2013). Bhāvamiśra argued that the *phiraṅgiṇī* (European women, a term also applied to Eurasians) were to blame for spreading the infection (Wujastyk 2015b). This seems surprising given the extremely small number of European women or even Eurasians in India at the time.[4]

[4] As Subrahmanyam (2012) notes, the numbers of Portuguese women in Asia represented a tiny fraction of the total number of settlers, despite some occasional efforts by the Portuguese state to bolster their numbers by despatching female orphans to the Indies. Portuguese men in Asia usually married local women, but their descendants sometimes retained elements of Portuguese identity.

As Wujastyk (2013) notes, Bhāvamiśra classified the Frankish disease as *āgantu* (invasive). This was one of three classes of disease, and it occurs in much earlier Ayurvedic texts, including the foundational work *Carakasaṃhitā* (c. first century CE). It is associated with the idea of contagion from at least the time of the *Suśrutasaṃhitā* (c. third century CE) (Wujastyk 2013). While this classification meant that the origin of the disease came from outside the body, it also affected the basic bodily constituents (called *doṣas* in Ayurvedic texts and usually translated as 'humours'). Bhāvamiśra (1998, I, 175) also classified syphilis as arising from *vāta* or 'wind', one of these humours. This classification was also applied to other venereal diseases, as well as other conditions, including insanity, epilepsy and bodily aches (ibid., II, 588). Bhāvamiśra's account of the Frankish disease was rare for its time but seems to have been widely circulated, and several other Sanskrit works dated to around the seventeenth century describe the *phiraṅga roga* (Wujastyk 2013).

The Dutch traveller Jan Huygen van Linschoten (1563–1611) described the Frankish disease in his account of his travels in Portuguese India during the 1580s. He claimed that in Goa venereal disease was not considered shameful and pointed to cases in which people boasted of having contracted the malady two or three times (Linschoten 1885, II, 107.). It is possible that by the late sixteenth century in Asia, as in Europe, the disease had reached an endemic steady state, meaning that it was common but less deadly. However, a useful counterbalance to Linschoten's assessments of attitudes towards the Frankish disease in India can be found in *Ardhakathanak* ('A half story'), the autobiography of the Jain poet Banārasīdāsa (1586–1641). In 1602, Banārasīdāsa experienced what he described as 'a morbid attack of *vāta*',[5] during which he felt not only the physical pain of the boils, blisters and aches but also great shame: men shunned him, his parents wept and wailed, and his new wife was prevented by her parents from returning home with him. The idea that the disease was a punishment for personal sin also emerges clearly from this account (Banārasīdāsa 2010, 79).

In Sri Lanka, the Sinhala medical compendium *Yogaratnakara* describes a disease known as *parangi* (meaning literally 'Portuguese') (Attygale 1917; de Silva and Gomez 1994).[6] The author of the *Yogaratnakara* is unknown, but according to some reports it was produced in the court of

[5] As noted above, *vāta* ('wind') is associated with venereal disease in Ayurvedic medicine.

[6] The Sinhala *Yogaratnakara* is not to be confused with the Sanskrit text of the same name.

King Buwanekabahu VII, who ruled Kotte from 1521 to 1550 (Uragoda 1987, 265). According to this work, the new disease had been brought to the island by the African slaves shipped there by the Portuguese (Attygale 1917).

In China the new disease that appeared in around 1505 was sometimes known as 'Guangdong sores', a reference to its apparent arrival via the port known to Europeans as Canton. Another common name was *yang-mei chuang* (red bayberry sores). It was discussed in detail by some sixteenth-century medical writers. Wang Ji (1463–1539), a practitioner of medicine in Qimen in modern-day Anhui province and the author of the *Shishan yi'an* ('Stone Mountain medical case histories'), encountered several patients suffering from *yangmei chuang* (Grant 2003). Wang Ji considered the disease to be the result of an accumulation of humidity and heat in the muscles and interstices, resulting in carbuncles, swellings and spasms (Schonebaum 2016, 129). Rather than identifying sex as a source of contagion, he thought instead that sexual licentiousness could interrupt the body's natural functioning, allowing the humidity and heat to dominate.

Li Shizhen (c. 1518–1593) elaborated on Wang Ji's understanding of the disease in his *Bencao Gangmu* ('Systematic materia medica'). Li was from a family of healers in Qizhou (now Qichun in Hubei province), but he lived in Beijing and spent his early years travelling through southern China before returning to his birthplace to write the voluminous *Bencao Gangmu*, a work that would become extremely influential in and beyond China after its posthumous publication in 1596 (Li Shizhen 2003; Nappi 2009). Li agreed with Wang Ji that the *Yangmei chuang* were the result of excessive bodily humidity and heat. He suggested the hot weather in the south of China as a cause, added to which, he writes, people indulge in hot food and 'run wild in sex'. However, Li also identified sexual contact as a means of *chuanran* ('contagion'). From the late Ming period onwards, the concept of *chuanran* gradually became more important in Chinese medical thought, in part as a result of the new disease (Schonebaum 2016). Chinese medical writers like Li often linked environment and contagion, believing that people could embody and transmit to others the effects of unhealthy climatic conditions such as miasmas (ibid.).

According to the *Gekkai-Roku*, a medical essay written in the Muromachi period, the new disease appeared in Japan in 1512. Early names for the condition in Japan included *tōkasa* (or *tō-mo*), *liukiu-kasa* and *namban-kasa* (Dohi 1923, 49), the latter indicating its foreign origins. In his work *Shittei Mondō* of 1563/4, the celebrated physician Manase Dōsan (1507–1594) records treating the condition (Dohi 1923; Kehoe 1992). Dōsan was a Sinophile who is said to have largely taken on Ming understandings of the new disease. Later Japanese works continued to draw on Chinese works that describe the disease, including Li Shizhen's *Bencao Gangmu*. They were also influenced by European understandings of venereal disease, based on translations from a number of Dutch works (Kehoe 1992).

By the seventeenth century, in both the Chinese and Japanese medical literature, the spread of the disease was linked to prostitutes (Schonebaum 2016; Kehoe 1992). Indeed, the confinement of prostitutes to segregated licensed parts of cities had begun even earlier, by the sixteenth century. Also during that century, Europeans visiting China and Japan were limited to particular trading points, and, in Japan, interactions between foreigners and Japanese women were limited to licensed prostitutes (Kehoe 1992).

In some parts of the IOW, notably sub-Saharan Africa and Southeast Asia, references to the Frankish disease seem rare or absent. Leo Africanus mentioned that the disease was not present in sub-Saharan Africa (in fact, he claimed that just going there was enough to cure a sufferer). This seems surprising given that Africa was integrated into shipping networks across the Atlantic and Indian Oceans, as well as the Mediterranean. The apparent lack of an epidemic of venereal disease there might perhaps be explained by the prior presence of yaws, another form of treponemal disease, that is thought to inhibit the spread of syphilis (populations that have suffered from yaws over a long period are likely to be resistant to infection with syphilis). The reference in the sixteenth-century Sinhala *Yogaratnakara* to the *parangi* disease that arrived from East Africa does suggest the continent as a source of epidemic disease, but it is unclear if the condition was venereal. In fact, later cases of *parangi* have been identified with yaws rather than syphilis (de Silva and Gomez 1994).

The prior presence of a form of treponemal disease has also been proposed for the apparent lack of a major epidemic of the new disease in early modern Southeast Asia. While reports from the 1580s mention *bubas* (literally 'pustules'), a condition for which a dedicated hospital existed in Manila by the 1590s, references to the disease do not necessarily point to venereal transmission (Newson 2009). The VOC physician Jacob de Bondt, or Jacobus Bontius (1592–1631), described an endemic 'pox' common in the island of Amboyna and especially in the Moluccas, which he said was very much like the *lues venerea* (venereal 'plague'), but was not sexually transmitted (Bontius 1769, 82–84). However, the *lues venerea* itself is not mentioned in Bontius' account of epidemic diseases in the region.

It appears from this survey that many medical authorities across the IOW associated the arrival of Europeans with the spread of the Frankish disease. The date at which the new disease had appeared was less clear, and the evidence suggests that the increasing intrusion of Europeans into societies across the IOW was retrospectively blamed for a range of illnesses before and after the probable appearance of the Great Pox in Europe (although notably in texts composed after the 1490s). Disease was often described as arriving in times of conflict, and it was thought to accompany the migration of groups of people like the Jews in North Africa or East African slaves in Sri Lanka. Most writers made a connection between the Frankish disease and sexual relations. Many medical cultures had some sense of 'contagion', which was applied to the new disease and sometimes gained wider recognition as a result. However, these ideas about contagion usually co-existed with humoral explanations for disease. Environmental explanations connected with these humoral ideas included miasmas, heat and humidity.

Some ideas expressed by medical writers from the IOW seem surprising. For example, what should we make of the claim of Leo Africanus that Jewish women were key carriers of the Frankish disease or Bhāvamiśra's laying the blame for it at the feet of European women? After all, foreigners who entered the region and had sexual relations with locals—usually soldiers or sailors followed by settlers—were much more likely to be male. We might speculate that this surprising claim was based on a greater readiness among the generally patriarchal societies of this period to countenance the idea of local men having sex with foreign women rather than vice versa, as well as a sense of women as reservoirs of disease based on the anatomy of the sexual organs (Kehoe 1992; Dohi 1923; Schonebaum 2016).

The attribution of the spread of the Frankish disease to Europeans continued beyond the point at which they could realistically be blamed for the majority of the contagion, given the evidence that the disease had become widespread across the region by the mid-sixteenth century. This tendency probably reflects a wider concern about the social and political disruption caused by the presence of Europeans, expressed through the language of disease. The idea that venereal disease was not considered shameful outside Europe seems incorrect, and, especially in parts of Asia, the disease gradually began to be regarded as the problem of social 'outsiders', such as inferior classes or prostitutes.

Treatments

Can the IOW reasonably be regarded as a distinct zone for medical and healing practices, as well as for the communication and conceptualization of disease? In this section I shall address this question by looking at some treatments for the Frankish disease.

In each part of the world that was affected by the new disease, treatments were improvised, often based on analogy with known diseases that were considered similar and the humoral profiles assigned to the disease. Drugs used to treat the Frankish disease tended to be those conventionally used in treating skin conditions or inflammation. For example, aloes appear in the prescriptions given by ʿImād ad-Dīn (Elgood 1931). Interestingly, drugs regarded as aphrodisiacs (including both China root and mercury, discussed below) were popular treatments. People also turned to general purgatives such as myrobalans, used in both India and Persia, and substances considered universal panaceas such as the medicinal tree neem (*Azadirachta indica*), used in both South Asia and parts of Africa (Arnold 1993; Bhāvamiśra 1998; Elgood 1931; Wujastyk 2013, 1049).

A few cures for the Frankish disease became truly global: these included the wood, bark or gum of guaiacum, the roots of smilax and various preparations of mercury. In the next sections of this chapter, I shall discuss the spread and reception of these three remedies across the early modern world and consider how they managed to travel across diverse medical cultures.

Guaiacum

Guaiacum (from the Spanish *guayacán*) is a common term for five evergreen shrubs or trees of the genus growing in Central and South America and what is now Florida. The wood, bark or resin of either *Guaiacum officinale* or *Guaiacum sanctum* is most commonly used in medicine. Early accounts note that the use of guaiacum was communicated by a slave or servant on Hispaniola (modern Haiti or the Dominican Republic) to a European suffering with the Pox. Reports and shipments of the drug reached Spain between 1506 and 1515, and by 1517 the remedy was in common use throughout Europe. The debate over its effectiveness continued for several centuries. Early proponents of guaiacum included the humanist writer Ulrich von Hutten (1488–1523). A vehement detractor was the physician and alchemist Paracelsus (1493–1541). The Fugger family of financiers had a virtual monopoly of sales of the drug in Europe (Munger 1949).

Whether guaiacum actually was used in the pre-Columbian Americas as a medicine remains debatable. Recipes clearly mentioning the plant in Aztec and Mayan literature before European contact have yet to be discovered (ibid.). However, guaiacum soon appeared as a staple item in Spanish pharmacies in Mexico (Paula de Vos, personal communication), and it is described in the early eighteenth-century writings of the Mexican physician Nicolás Joseph de Torres (Torres c. 1720).

While guaiacum does not seem to be mentioned in any of the medical treatises from the IOW, a few references suggest that it was known and traded there. Linschoten (1885 II, 107) wrote that, on its first appearance in the East Indies, the wood was 'weighed against gold'. Similarly, the physician Garcia da Orta brought stocks of guaiacum with him from Portugal to Goa in 1634, indicating that he was aware of a market for the drug in Asia (Orta 1913, 381). The French traveller André Thevet (died 1590) (Thevet 1575, VIII, 137) noted the availability of guaiacum on the southern Arabian coast. These references probably indicate a trade carried on by Portuguese merchants. However, all these authors argue that guaiacum fell from popularity with the arrival of the so-called China root from 1535 (see discussion of *Smilax* below).

Guaiacum continued to be used in Europe and America, and it still has medical applications, notably as a cough medicine. It also held a place in colonial medical practice in the IOW. For example, Bontius (1769, 82–84) mentioned using guaiacum to treat the 'Amboyna pox'. A British customs order dated 1797 restricted its export from Britain to the British colonies in the east in order to ensure its availability to physicians in those colonies (Customs-House 1797). And in the mid-nineteenth century, Alexander Gibson, an English surgeon and botanist, suggested transplanting guaiacum to India, along with other American medicinal plants (Bombay Castle 1839, 1847). Although it was clearly available in the IOW, guaiacum does not seem to have gained widespread popularity within local medical practice.

Smilax

The new treatment that had so quickly replaced guaiacum, according to Linschoten, Orta and Thevet, was the root of the small climbing woody plants *Smilax china* or *Smilax glabra*. Termed *tu fu ling* in Chinese and *dobukuryo* or *sankirai* in Japanese, the drug was otherwise widely known as 'China root', a name that probably originated from the Persian word *chub-i-chini* (Trambaiolo 2015; Wujastyk 2015a). This remedy for the Frankish disease came into use in the southern region of China between the arrival of the new disease in 1505 and around 1511, when, according to some reports, China root first arrived in Malacca. From 1535 onwards, China root spread quickly across the world, often travelling along the same routes along which the new disease had spread: army camps and trans-IOW trading networks.[7]

China root came to be popularly cited as a cure in medical literature across the IOW, appearing in the works of many of the authors mentioned above. In Persia 'Imād ad-Dīn dedicated a treatise to it (Elgood 1931, 1951). In Japan it was used by Manase Dōsan in the 1560s (Dohi 1923). It appears in the mid-sixteenth-century treatise of Bhāvamiśra (1998), and Linschoten (1885) noted that it was cheap and readily available in western Indian bazaars by the 1580s. In that decade in Cairo, the Venetian physician and botanist Prosper Alpini (1553–1617), who was then residing in the city, noted its popularity for a number of complaints (Alpini 1980). China root maintained its popularity in the IOW. For example, in a brief manuscript account, Sulayman b. Ali al-Mundhiri, who was an Omani lawyer at the court of the late nineteenth-century Ālbūsa'īdī Sultanate of Zanzibar, describes China root (*al-šubšīnī*) and the plant it comes from, gives instructions for its preparation and describes its use against the Frankish disease and other conditions (Declich 2004).[8] China root also remained an important remedy for venereal disease across Asia and the Middle East up until the nineteenth century, and it is still used for this and many other purposes in South Asia and East Asia.

China root also reached the Americas, where Francisco de Mendoza (active c. 1524–1563), son of the first viceroy of New Spain, Antonio de Mendoza, tried transplanting it to New Spain in 1558. While he seems to have had some success, his efforts were soon rendered needless by the discovery that several species of the genus *Smilax* also grew wild in the

[7] I have discussed the China root at length elsewhere (Winterbottom 2015).

[8] The manuscript is ZA 2/4, Zanzibar National Archives, and the account is on ff. 90–91. I am very grateful to Lorenzo Declich for sharing both his photograph of the document and his own translation of it. The name that is given to the China root here, *al-šubšīnī*, is clearly derived from the Persian *chub-i-chini* (the Arabic name used elsewhere is *labana*). The author lists the names *šajar, ḥabb ifranjī, mubārak* (the last meaning literally 'the blessed') as being in use in the Islamic world for the Frankish disease.

Americas. These plants came to be known as sarsaparilla (from Spanish *zarzaparrilla*; small, brambled vine). As with guaiacum, the medical benefits of sarsaparilla were apparently communicated by natives of the region, in this case those of New Spain. Subsequent discoveries of sarsaparilla in Brazil, Virginia and Jamaica prompted the Dutch and English to break into the valuable trade in anti-syphilitics. By the mid-nineteenth century, the Jamaican sarsaparilla (*Smilax aspera*) had become an extremely valuable commodity and was produced on a large scale on plantations.

American sarsaparilla largely succeeded in displacing the demand for China root in Europe and the Atlantic world.[9] By the mid-nineteenth century, American sarsaparilla was, at considerable expense, being imported into India for the use of colonial physicians, despite the availability of China root and local substitutes (Bombay Castle 1847). In fact, even the *Smilax* species known as sarsaparilla in America grew wild in parts of the IOW, notably in Southeast Asia. For example, the soldier and naturalist Georgius Everhardus Rumphius (1627–1702) noted that *Smilax aspera* was present in Amboyna and identified it with samples he had received from the Americas (Rumphius 2011, iv, 491–500). Some medical authors from the IOW also commented on the trade in American sarsaparilla. Elgood (1951, 53–54) notes the chapter devoted to the subject by Hakim Mohamed Hashim bin Mohammad Tahir, a Persian author of the late Safavid period. Renaud and Colin (1935) translated sections from the work of an otherwise unknown author, aṣ-Ṣiqillī, possibly a Tunisian, on the relative virtues of the roots brought by Europeans from 'the two places they call the Indies'. In general, however, China root continued to be preferred to sarsaparilla as a medicine in the IOW.

Mercury

Guaiacum and smilax were new drugs on the international stage in early modernity, and it is therefore relatively straightforward to trace where they were originally used to treat the new disease. Mercury, the third global remedy, is more complex, as it was known throughout the ancient world. Early modern exchanges of information about the medical uses of mercury were multidirectional, and it is sometimes hard to disentangle the webs of exchange to point to the origins of particular medical applications (Wujastyk 2015a).

[9] For details of this process, see Winterbottom (2015).

Mercury and cinnabar (the red mineral ore of mercury, HgS) were probably first medically used in China, based on their presence in burial sites from as early as the second millennium BCE (Trambaiolo 2015). Concern over the toxic properties of mercury seems to have fluctuated over time. The Greek medical authors Pedanius Dioscorides (c. 40–c. 90 CE) and Galen of Pergamum (c. 129 CE–210 CE) both regarded mercury as a poison, and the Byzantine physician Paul of Aegina (c. 625–c. 690 CE) agreed while noting that it was used internally as a cure by some of his contemporaries (Bachour 2015).

While mercury was also recognized as a poison by Indian medical writers, it came to be accepted over time as suitable for internal as well as external use. Important to the increasing acceptance of mercurial medicines was the development of various methods intended to purify it or to remove its toxicity. These involved sublimation and calcination (to yield mercury chloride and corrosive sublimate) and a process known as 'killing' mercury (originally referring to a chemical procedure, this also came to denote mixing mercury derivatives with other ingredients) (Bachour 2015; Wujastyk 2015b). These techniques were described in Indian alchemical works by the ninth century and became known in Persia and in the Arabic medical literature from the time of the well-known Persian philosopher and alchemist Abū Bakr Muḥammad ibn Zakariyyāʾ al-Rāzī (in Latin, Rhazes, c. 865–925/935 CE) and Ibn Sīnā (Bachour 2015; Wujastyk 2015b). While 'killed' mercury was taken internally in South Asia for various conditions from the thirteenth century onwards, application in the form of an ointment to treat skin disorders remained more common in the Middle East. In the Western Islamic world, liquid mercury also found medical applications. During the Middle Ages, these external uses became known in Europe through translations from Arabic into Latin (Bachour 2015).

Given the existing uses of mercury to treat skin conditions and venereal diseases, it was natural that mercurial therapies were applied to the Frankish disease in Persia and South Asia. In East Asia, the medical use of mercury had apparently fallen out of popular use, but it saw a renaissance in the sixteenth century (Trambaiolo 2015). Mercury chloride (calomel) was swiftly applied to the treatment of syphilis in both China and Japan, as described by Li Shizhen (2003) and Manase Dōsan (Dohi 1923).

The first European application of mercurial pastes to the legs to treat the Pox is mentioned in Konrad Schellig's work of c. 1496.[10] According to legend, mercury pills were introduced to Europe by the Turkish corsair Khayr al-Dīn Pāshā (1466–1545), known as Barbarossa, who gave them to the King of France, Francis I (ruled 1515–1547) (Bachour 2015). While the pills may have in fact arrived more gradually, the story of a new medicine appearing and being tested in a courtly setting mirrors similar stories about China root, which was famously tried by the Holy Roman Emperor Charles V (who ruled as Emperor 1519–1556 and was also Charles I of Spain). Charles V's use of the root apparently prompted the famed anatomist Andreas Vesalius to write his epistle[11] on the subject (Garrison 2015).

Bahāʾ ad-Dawla mentioned a pill containing a small amount of mercury in 1501, and when ʿImād ad-Dīn later gave a more detailed account of it, he described it as well known in India, Constantinople and Europe. The pill seems likely to have been adopted from the east rather than the west, possibly by means of the circulation at the time of numerous Persian physicians through the Mughal and regional courts of India. Yūsufi mentioned fumigation with mercury in 1511, before its first mention in the work of European physicians, perhaps indicating the Persian origin of this particular method of treatment. Al-Anṭākī's work describes pills made from mercury and arsenic that were ordered from Venice, as well as preparations that seem to have originated from Persia or India (Bachour 2015; Thomann 2015).

Mercurial preparations quickly spread back from Europe to parts of the IOW. The internal use of mercury might have been encouraged by the growing popularity of the work of Paracelsus and his followers in Europe and the Middle East (ibid.). European preparations of mercury are mentioned by ʿImād ad-Dīn, although he cautions against one of them, as he considers it dangerous (Elgood 1931). In Japan, mercury chloride drugs made to both Chinese and European specifications were being imported by the eighteenth century (Trambaiolo 2015). Similarly, in Tibet, a range of uses of mercury were reported from the eighteenth century onwards,

[10]This work is known variously as *Consilium in morbum gallicum*, *Consilium in pustulas malas* and *De morbo gallico* (Goldwater 1972, 217). Schellig is otherwise little known.

[11]Andreas Vesalius, *Epistola rationem modum que propinandi radicis Chynae decocti, quo nuper inuictissimus Carolus V. Imperator usus est, pertractans*, Basil: Johannes Oporinus, 1546.

some apparently having been adopted from Chinese, Indian and European sources, while others built on classical medical writings from around the twelfth century that describe the use of mercury to treat venereal disease (Gerke 2015). In India, the use of mercury was championed by colonial physicians. For example, in 1808 a treatise on the use of mercury to treat venereal disease composed by an East India Company Surgeon, Dr Ingledew of Mysore, was translated into two South Indian languages to encourage its adoption by local physicians, who had apparently expressed misgivings about its toxicity (Madras Military Board 1808).

To summarize, the various methods of preparing drugs using mercury seem to have originated in Asia and spread across the world. While the poisonous effects of mercury were widely recognized across the world from an early stage, the amounts of mercury in these preparations gradually increased as they continued to circulate during the early modern period. One Japanese explanation was that the poison of a virulent disease could only be expelled by a drug that had violent effects (Trambaiolo 2015). We might also speculate that greater circulation of remedies and openness to 'exotic' preparations in the early modern world led to less caution all around.

What propelled certain remedies for syphilis out of the local sphere and into global circulation? In the case of both guaiacum and the *Smilax* species, part of the attraction was novelty: they were new remedies for a new disease. Their global accessibility was facilitated by international networks of traders—notably the Portuguese—who helped supply local markets. Mercury was seen as a powerful medical substance with sexual associations and had the advantage of being widely available across the world. Demonstrations of drugs in royal courts and army camps helped them to move between different medical cultures which nonetheless shared a basic understanding of health as rooted in a balance of key bodily substances. Finally, the championing of the new drugs by well-respected medical authors raised their profile. The division that emerged between the use of guaiacum and sarsaparilla in the Atlantic world and China root in the IOW suggests that these may reasonably be regarded as distinct zones for the spread of treatments as well as diseases.

Re-imagining the Frankish Disease

Studies of colonial rule in the nineteenth and early twentieth centuries have noted the grave concern that colonial authorities expressed about the disease that was by then called syphilis by Europeans. Medical reports from this period described very high rates of syphilis, allegedly affecting sometimes up to 85–90 per cent of the indigenous populations of Morocco (Amster 2013); eastern, central and southern Africa (Vaughan 1992); and India (Arnold 1993; Levine 1994; Wald 2014). As historians have noted, these figures probably covered other skin conditions, as well as venereal complaints such as gonorrhoea and other treponemal diseases, notably yaws, which was only distinguished from syphilis from the beginning of the twentieth century.

Draconian methods of controlling syphilis begun in Europe were often exported to the colonies, where they remained in force long after they were considered unacceptable in the colonial metropole. Measures taken in India from around 1800 to prevent the further spread of the disease in the British army, on which it was said to be inflicting considerable damage, included the use of 'lock' hospitals to treat Indian women who were believed to be affected (ibid.). The Cantonments Act of 1864 regulated the sex trade within military towns, and the Indian Contagious Diseases Act of 1868 provided for the supervision, registration and medical inspection of female Indian prostitutes in all major cities and ports. The second act also established a distinction between prostitutes who were intended to provide sexual services solely to European men and those meant to provide them only to Indian men (Levine 1994, 581, 586).

Similar measures were taken in Sri Lanka (de Silva and Gomez 1994), Hong Kong and Aden, where, as late as the early 1900s, the British colonial authorities and army medical officials collaborated to require both 'public' and 'private' prostitutes to submit to weekly examinations. Prostitutes who were thought to have infected British soldiers were required to attend hospital until cured, on pain of banishment from the area (Aden Protectorate 1910–1932). Even after such measures were outlawed, the colonial state continued to keep a close watch on the activities of prostitutes, going so far as to draw maps of their zones of activity (Aden Protectorate 1932–1936).

As Megan Vaughan (1992) has shown, colonial concern over syphilis epidemics was expressed in eastern, central and southern Africa during the early twentieth century; in Rhodesia, venereal disease was cited as a justification for racial segregation. By the 1950s, the understanding of yaws as a separate, non-venereal condition had finally emerged, along with recognition that yaws and syphilis might be mutually exclusive. In fact, it was even speculated that the treatment of yaws had made way for the spread of syphilis (ibid.). This suggestion from the late colonial period seems to support Leo Africanus' early claim that sub-Saharan Africa was free from the Frankish disease.

The argument historians often make for the modern colonial period, that extreme concern about sexually transmitted diseases arose in a large part from the uneasy co-existence of Europeans and locals, has rarely been applied to the early modern period. Equally, historians have seldom considered whether there really was another epidemic of the Frankish disease in the IOW during the late nineteenth and early twentieth centuries. In this regard, it might be worth remembering that this was another period in which European activity—and particularly military presence—intensified and reached new parts of the region (Jackson 2011). As in the earlier period, armies and navies composed of mobile single men are likely to have provided ideal agents for the spread of venereal disease.

Conclusions

"[F]rom the earliest chroniclers, the one constant feature of the disease was that it did not arise among the group to which the author belonged" (Arrizabalaga et al. 1997, 12). This statement about the Great Pox in Europe could be applied equally to the Frankish disease in the IOW. Arrizabalaga et al. (1997) argue that one reason that the thesis of an American origin gained ground in Europe was that the natives of the Americas were the ultimate 'others' in the early modern world, having been unknown until Columbus's voyages. In the IOW, there seems to have been almost a consensus that the disease was associated with Europeans. This might perhaps be connected with the disruptive effects of the violent intrusions of Europeans into societies around the region, as well as the observation some authors made that the disease had originated in the west.

Arrizabalaga et al. (1997, 12) also note that the Great Pox stimulated debate over the nature of disease. Similarly, the Frankish disease prompted some changes in medical thinking in the IOW. While humoral understandings of the body that explained disease as a result of imbalances of bodily substances remained paramount (at least in medical writings), ideas about contagion became increasingly important. Accompanying this idea was the sense that close association with certain groups of people, whether foreigners or prostitutes, could be dangerous. In some societies, especially in East Asia, there were some governmental efforts to control such groups, for example, by limiting them to particular geographical areas.

By the nineteenth century, the early perception of the Frankish disease as something that came from Europe had been reversed, at least in the minds of the colonists, who now portrayed the natives of the IOW as the carriers of venereal diseases that were more virulent than the forms that existed in Europe (Levine 1994). Colonial regimes took efforts to limit sexual contacts between different social groups to an extreme. The colonial perception that local prostitutes were to blame for venereal disease seems to have had an enduring effect in deflecting attention from European armies as the possible carriers of disease in the nineteenth century.

The early cures for syphilis—notably guaiacum, China root, sarsaparilla and mercurial preparations—came from all over the early modern world. These cures were understood and applied in fairly similar ways across different medical cultures, although different explanations were given for their effectiveness. Thus, while the IOW can be regarded as a 'world' in the sense of shared understandings of the disease and preferences for particular drugs in treating it, it was not a world apart. Physicians and medical writers from the region participated in global debates about the causation of disease and the source of healing. The preference for China root in the IOW and for sarsaparilla in the Atlantic world and among some later colonial physicians had more to do with geopolitical allegiances (a preference for American or Chinese products) than fundamental differences in medical thought. This situation remained effectively unchanged up until the early twentieth-century identification of *T. pallidum* and the development of first the arsenic-based drug Salvarsan and then penicillin as effective treatments for syphilis.

References

Aden Protectorate. 1910–1932. *File 9/2 Prostitutes: Venereal Disease*. British Library IOR/R/20/A/1285.

———. 1932–1936. *File 66/2 Prostitutes*. British Library, IOR R/20/A/2213.

Alpini, Prosper. 1980. *La Médecine des Egyptiens*. Trans. R. de Fenoyl. Cairo: Institut Français d'Archéologie Orientale.

Amster, Ellen. 2013. *Medicine and the Saints: Science, Islam, and the Colonial Encounter in Morocco, 1877–1956*. Austin, TX: University of Texas Press.

Arnold, David. 1991. The Indian Ocean as a Disease Zone, 1500–1950. *South Asia: Journal of South Asian Studies* 14 (2): 1–21.

———. 1993. Sexually Transmitted Diseases in Nineteenth- and Twentieth-Century India. *Genitourinary Medicine* 69: 3–8.

Arrizabalaga, Jon. 2002. Problematizing Retrospective Diagnosis in the History of Disease. *Asclepio* 54 (1): 51–70.

Arrizabalaga, Jon, John Henderson, and R.K. French. 1997. *The Great Pox: The French Disease in Renaissance Europe*. New Haven, CT: Yale University Press.

Attygale, John. 1917. *Sinhalese Materia Medica*. Colombo: Colombo Apothecaries.

Bachour, Natalia. 2015. Healing with Mercury: The Uses of Mercury in Arabic Medical Literature. *Asiatische Studien* 69 (4): 831–866.

Banārasīdāsa. 2010. *Ardhakathanak: A Half Story*. Trans. Rohini Chowdhury. London: Penguin.

Bhāvamiśra. 1998. *Bhāvaprakasa of Bhāvamiśra: Text, English Translation, Notes, Appendices and Index*. 2 vols. Trans. and Ed. K.R. Srikantha Murthy. Varanasi: Krishnadas Academy.

Bombay Castle, General Department. 1839, October 23. *No. 2938, Copy of Letter from Alexander Gibson, Dated 30 September 1839*. British Library IOR P/348/5.

Bombay Castle. 1847, August 1. *Nos. 4028-4030, Letter of Alexander Gibson, Dated 8 February 1847*. British Library IOR P/349/46.

Bontius, Jacobus. 1769. *An Account of the Diseases, Natural History, and Medicines of the East Indies: Translation with Annotations of a Physician*. London: T. Noteman.

Conrad, Lawrence I. 1992. Epidemic Disease in Formal and Popular Thought in early Islamic Society. In *Epidemics and Ideas: Essays on the Historical Perception of Pestilence*, ed. Terence Ranger and Paul Slack, 77–100. New York: Cambridge University Press.

Cooter, Roger, and Claudia Stein. 2013. *Writing History in the Age of Biomedicine*. New Haven, CT: Yale University Press.

Crosby, Alfred W. 1969. The Early History of Syphilis: A Reappraisal. *American Anthropologist* 71: 218–228.

———. 1972. *The Columbian Exchange: Biological and Cultural Consequences of 1492*. Westport, CT: Greenwood Pub.

Customs-House, London. 1797. *Customs Notice and Quarantine Order*. Welcome Trust WMS 30 (Shelfmark MS. 8396).

De Silva, P.A., and M.G. Gomez. 1994. The History of Venereal Disease and Yaws (Parangi) in Sri Lanka (Ceylon). *Genitourinary Medicine* 70 (5): 349–354.

Declich, Lorenzo. 2004. *The Mundiris of Zanzibar: Three Generations of Ibād 'ulamā' at the Sultan's Court (1800–1900)*. Rome: Aracne.

Dohi, Keizo. 1923. *Beiträge zur Geschichte der Syphilis*. Tokyo: Nankodo.

Elgood, C. 1931. Translation of a Persian Monograph on Syphilis Entitled Risála-i-Átishak by Imád-ul-Din Maḥmúd bin Mascúd bin Maḥmúd-ul-Ṭabíb. *Annals of Medical History* 3: 465–486.

Elgood, Cyril. 1951. *A Medical History of Persia, and the Eastern Caliphate, from the Earliest Times until the Year A.D. 1932*. Cambridge: Cambridge University Press.

Épaulard, A., ed. and trans. 1960. *Leon l'africain: description de l'Afrique*. Paris: Librarie d'Amérique et d'Orient Adrien-Maisonneuve.

Garrison, Daniel H. 2015. *Vesalius, the China Root Epistle: A New Translation and Critical Edition*. Cambridge: Cambridge University Press.

Gerke, Barbara. 2015. The Poison of Touch: Tracing Mercurial Treatments of Venereal Diseases in Tibet. *Social History of Medicine* 28 (3): 532–554.

Goldwater, Leonard J. 1972. *Mercury: A History of Quicksilver*. Baltimore, MD: York Press.

Grant, Joanna. 2003. *A Chinese Physician: Wang Ji and the 'Stone Mountain Medical Case Histories'*. London: Routledge Curzon.

Harper, K.N., et al. 2011. The Origin and Antiquity of Syphilis Revisited: An Appraisal of Old World Pre-Columbian Evidence for Treponemal Infection. *American Journal of Physical Anthropology, Suppl. Yearbook of Physical Anthropology* 53: 99–133.

Harper, Kristin N, Molly K. Zuckerman, and George J. Armelagos. 2014. Syphilis: Then and Now. *The Scientist*. http://www.the-scientist.com/?articles.view/articleNo/38985/title/Syphilis%2D%2DThen-and-Now/.

Jackson, Ashley. 2011. Britain in the Indian Ocean Region. *Journal of the Indian Ocean Region* 7 (2): 145–160.

Kehoe, Jean. 1992. *Imperial Medicine, Venereal Disease and Social Relations in Japan, 1542–1858*. Ph.D. dissertation, Victoria University of Washington.

Levine, Philippa. 1994. Venereal Disease, Prostitution, and the Politics of Empire: The Case of British India. *Journal of the History of Sexuality* 4 (4): 579–602.

Li, Shizhen. 2003. *Compendium of Materia Medica: Bencao Gangmu*. Trans. Xiwen Luo. Beijing: Foreign Languages Press.

Linschoten, John Huyghen van. 1885, January. *The Voyage of John Huyghen Van Linschoten to the East Indies: From the Old English Translation of 1598: The First Book, Containing His Description of the East, in Two Volumes*. Eds. A.C. Burnell, Pieter A. Tiele, and William Phillip. London: Hakluyt Society.

Madras Military Board. 1808, November 18. *India Office Records*. British Library P/256/40: 10940–46.

Munger, Robert S. 1949. Guaiacum, the Holy Wood from the New World. *Journal of the History of Medicine and Allied Sciences* 4: 196–229.

Nappi, Carla S. 2009. *The Monkey and the Inkpot: Natural History and Its Transformations in Early Modern China*. Cambridge, MA: Harvard University Press.

Newson, Linda A. 2009. *Conquest and Pestilence in the Early Spanish Philippines*. Honolulu: University of Hawai'i Press.

Nutton, Vivian. 1990. The Reception of Fracastoro's Theory of Contagion: The Seed that Fell among Thorns? *Osiris*, 2nd series, 6: 196–234.

da Orta, Garcia. 1913. *Colloquies on the Simples and Plants of India*. Trans. Clements Markham. London: Henry Southern & Co.

Powell, Mary Lucas, and Della Collins Cook, eds. 2005. *The Myth of Syphilis: The Natural History of Treponematosis in North America*. Gainesville, FL: University Press of Florida.

Renaud, Henri-Paul-Joseph, and G.S. Colin. 1935. *Documents marocains pour servir à l'histoire du 'mal franc'*. Paris: Librairie Larose.

Rumphius, G.E. 2011. *The Ambonese Herbal*. 6 vols. Trans. and Ed. E.M. Beekman. New Haven, CT: Yale University Press.

Savage-Smith, Emilie. 1998. ʾEmād-al-Dīn Maḥmūd. *Encyclopaedia Iranica*, Vol. VIII, Fasc. 4. New York: Columbia University, Centre for Iranian Studies, pp. 381–382. http://www.iranicaonline.org/articles/emad-al-din-mahmud.

———. 2013. Were the Four Humours Fundamental to Medieval Islamic Practice? In *The Body in Balance: Humoral Medicines in Practice*, ed. Peregrine Horton and Elizabeth Hsu, 89–106. New York: Berghahn Books.

Schonebaum, A. 2016. *Novel Medicine: Healing, Literature, and Popular Knowledge in Early Modern China*. Seattle, WA: University of Washington Press.

Serjeant, Robert Bertram. 1965. Notice on Frankish Chancre (Syphilis) in Yemen, Egypt, and Persia. *Journal of Semitic Studies* 10: 241–252.

Stearns, J.K. 2011. *Infectious Ideas: Contagion in Pre-Modern Islamic and Christian Thought in the Western Mediterranean*. Baltimore, MD: Johns Hopkins University Press.

Subrahmanyam, Sanjay. 2012. *The Portuguese Empire in Asia, 1500–1700: A Political and Economic History*. Chichester: John Wiley & Sons.

Thevet, André. 1575. *La cosmographie universelle*. Paris: Guillaume Chandiere.

Thomann, Johannes. 2015. Early Persian Medical Works on Antisyphilitic Mercury Medicines. *Asiatische Studien – Études Asiatiques* 69 (4): 971–996.

Torres, Nicolás Joseph de. c. 1720. Virtudes del guaicán o palo santo. Unpublished notebook. Wellcome Trust MSS Americana 11.

Trambaiolo, Daniel. 2015. Antisyphilitic Mercury Drugs in Early Modern China and Japan. *Asiatische Studien – Études Asiatiques* 69 (4): 997–1016.

Uragoda, C.G. 1987. *History of Medicine in Sri Lanka from the Earliest Times to 1948*. Colombo: Sri Lanka Medical Association.

Varthema, Ludovico di. 1928. *The Itinerary of Ludovico di Varthema of Bologna from 1502 to 1508*. Ed. R.C. Temple. London: Argonaut Press.

Vaughan, Megan. 1992. Syphilis in Colonial East and Central Africa: The Social Construction of an Epidemic. In *Epidemics and Ideas: Essays on the Historical Perception of Pestilence*, ed. Terence Ranger and Paul Slack, 269–302. Cambridge: Cambridge University Press.

Wald, Erica. 2014. *Vice in the Barracks: Medicine, the Military and the Making of Colonial India, 1780–1868*. Basingstoke: Palgrave Macmillan.

Winterbottom, Anna E. 2015. Of the China Root: A Case Study of the Early Modern Circulation of Materia Medica. *Social History of Medicine* 28 (1): 22–44.

Wujastyk, Domanik. 2003. *The Roots of Ayurveda*. London: Penguin Books.
Wujastyk, Dagmar. 2013. Perfect Medicine: Mercury in Sanskrit Medical Literature. *Asian Medicine* 8 (1): 15–40.
———. 2015a. Histories of Mercury in Medicine Across Asia and Beyond. *Asiatische Studien – Études Asiatiques* 69 (4): 819–830.
———. 2015b. Mercury as an Antisyphilitic in Ayurvedic Medicine. *Asiatische Studien – Études Asiatiques* 69 (4): 1043–1067.

CHAPTER 4

Reconsidering the Early History of Leprosy in Light of Advances in Palaeopathology

Eric A. Strahorn

Introduction

Despite advances in DNA and molecular studies, relatively little is known about the early history of leprosy. Fanciful theories about disease transmission across the Indian Ocean world have contributed an element of empty speculation to the literature, and it is necessary to refocus on the actual evidence available. This chapter reviews the palaeopathological, archaeological and literary evidence and argues that DNA analysis has provided a great deal of new data on the history of leprosy, but the interpretation of this new evidence has proven to be problematic. Given the limitations of this rapidly developing DNA sequencing technology, there are still many unanswered questions about the role of the Indian Ocean world in the early history of leprosy. This chapter first defines leprosy, then evaluates the hypotheses for the early history of leprosy based on the literary and archaeological evidence, third analyses the claim that DNA evidence can fill some of

E. A. Strahorn (✉)
Florida Gulf Coast University, Fort Myers, FL, USA
e-mail: estraho@fgcu.edu

G. Campbell, E.-M. Knoll (eds.), *Disease Dispersion and Impact in the Indian Ocean World*, Palgrave Series in Indian Ocean World Studies, https://doi.org/10.1007/978-3-030-36264-5_4

the gaps in the literary and archaeological record and finally examines recent hypotheses that attempt to reconcile the literary, archaeological and DNA evidence for the early history of leprosy.

A Definition of Leprosy

Leprosy (sometimes termed "Hansen's disease") is currently defined as the disease that results from an infection with the bacterium *Mycobacterium leprae*, identified in 1873 by Norwegian physician Armauer Hansen (1841–1912). *M. leprae* is an obligate intracellular parasite for which humans are the only ubiquitous reservoir (it cannot be cultivated on an artificial medium or tissue). It is slow growing and in those affected can have an incubation period of three to ten years. The genome of *M. leprae* has been stable for at least the last 1000 years, but nearly half of its genes are nonfunctional. The exact route of transmission is thought to occur through nasal secretions (Britton 2017, 954–955; Cole and Singh 2012, 3; Gelber 2015; Virmond, Grzybowski and Virmond 2015, 3).

There are two main types of leprosy, lepromatous leprosy (LL) and tuberculoid leprosy (TT). Upon infection, the host immune response is responsible for the clinical features of the disease. LL, which is caused by a weak immunologic reaction, produces extensive skin lesions as well as damage to the eyes, nose, bones, testes, spleen, liver and adrenals. It can result in blindness, loss of extremities, and rhinomaxillary syndrome (leonine facies) which includes thickening of the skin, loss of eyebrows and eyelashes and erosion of the cartilage and nasomaxillary bones—resulting in a collapse of the nose (Britton 2017, 955–957; Gelber 2015; Virmond et al. 2015, 14). This is the type of leprosy that was "usually conjured in the medieval [European] imagination … and inspire[d] horror in the observer" (Zimmerman 2008, 560). TT is less severe due to a stronger immunologic reaction in the host. Affected areas include the nerves and skin, often with a small number of lesions that present with discolouration and loss of hair (Britton 2017, 956; Virmond et al. 2015, 17–18).

Hypotheses for the Early History of Leprosy

Before the first genotyping of *M. leprae* was accomplished in 2005, hypotheses relating to the early history of leprosy drew upon the fragmentary and contested literary and archaeological evidence available. One of the most influential hypotheses from the second half of the twentieth century, when archaeological evidence was available to supplement literary texts, was that the armies of Alexander the Great transmitted leprosy from India to the Mediterranean. One of the leading exponents of this hypothesis is Johannes Andersen, who considers that, unlike ancient Egyptian works, Indian texts such as the *Sushruta Samhita* reliably describe Hansen's disease several centuries before the Greek and Roman texts (Andersen 1969, 17–45), Browne (1975, 14) and Hulse (1975, 88). Andersen's timeline appeared to be confirmed following the discovery of second century BCE skeletons with rhinomaxillary syndrome at the Dakhleh Oasis, because it offered evidence that leprosy was present in Egypt after the time of Alexander's conquests (Roberts and Manchester 2005, 201; Miller and Nesbitt 2014, 10–11).

Other scholars question whether soldiers present in India for only a couple of years could have contracted leprosy. Thus Margaret Lloyd Davies and T.A. Lloyd Davies (1989, 623) propose that it was the soldiers' "10,000 Asian mistresses" who accompanied them to the Mediterranean who carried leprosy to Europe. However, there is no direct evidence to support their view. Keith Manchester (1984, 168) considers Andersen's hypothesis to be reasonable, but contends that, given the long history of caravans travelling between India and the Mediterranean, no specific date can be identified for the transmission of leprosy. Samuel Mark, on the basis of his analysis of several ancient texts including Kautilya's *Artaśāstra* or *Arthashastra* (Mark 2002, 302), argues that slave girls exported by ship from Mauryan India to Ptolemaic Egypt carried leprosy. However, with a few exceptions (e.g. Donoghue et al. 2015b, 251), his views have had little scholarly impact. This is because although the *Arthashastra* discusses both slavery and foreign trade, the earliest recension of the text dates to the first or second century AD and hence raises doubts about the accuracy of its descriptions of conditions during the Mauryan dynasty (322–187 BCE) and, moreover, does not mention the exportation of slaves from India (McClish 2012, 280–281; Olivelle 2013, 27–29).

Literary Evidence from Ancient China, India, Egypt and the Mediterranean

The early history of leprosy is poorly understood. It has typically involved the combing of written texts, works of art and other human artefacts for any hint of a reference to or description of the disease we now call leprosy. Prior to the twentieth century, medical and non-medical understandings of leprosy varied greatly across time and space. Hitherto, efforts to find any mention of leprosy in ancient texts have been beset with unanswerable questions and controversy with the result that the various theories about the transmission of the disease across the Indian Ocean world (IOW) are being revisited by scholars.

The discussion about the presence of leprosy in ancient China centres on three works, the *Analects* of Confucius, the medical text *Huangdi Neijing* (Wade-Giles romanisation, *Huang Ti Nei Ching*) and a legal document included in the *Ma-wang-tui* texts recovered from the third century BCE tomb of a government official. Scholars disagree as to whether or not ancient Chinese medical terms such as *li* and *dafeng* refer to Hansen's disease. The *Analects* is traditionally attributed to the philosopher Confucius (Kong Fuzi, 551–479 BCE) but is now thought to be the work of several authors dating from perhaps 479 to 249 BCE (Brooks and Brooks 1998, 205). Gwei-Djen Lu and Joseph Needham claim that a disciple of Confucius who suffered from a terrible affliction in fact had leprosy, even though the text does not describe the disease. They argue further that this interpretation is widely shared by Confucian scholars (Lu and Needham 1967, 229). However, Derk Bodde (1982, 10) considers such evidence to be "uncertain," and Angela Leung (2009, 81) suggests that Lu and Needham have engaged in "overinterpretation" and that there is nothing in the text to support the identification. Indeed, Needham and Lu (2000, 185) later admit that the text is too "ambiguous" to identify the disease cited with leprosy. There is also disagreement over *Ma-wang-tui* tomb texts, including a legal document that records the investigation of a magistrate into an individual accused of having *li*. Bodde (1982, 9) and Needham and Lu (2000, 184) believe that the description of the affliction is specific enough to identify leprosy, but Leung (2009, 4) counters that while the description is "suggestive," it is not enough to establish the presence of true leprosy in China during the third century BCE.

The identification of leprosy with the terms *li* and *dafeng* in the medical text, *Huangdi Neijing* (*Yellow Emperor's Inner Canon*), is also contested. This text is often dated to the second or first century BCE but in fact underwent numerous recensions between perhaps the second century BCE and eighth century CE (Unschuld 2003, ix). D.W. Beckett (1987, 494) declares that the book contains "the earliest truly recognisable description of leprosy," but John Lowe (1947, 56), Bodde (1982, 10) and Leung (2009, 18) all find the descriptions insufficient proof of leprosy. Indeed, Leung (ibid., 19) suggests that the compilers of the text may not have considered *li* and *dafeng* to be related and that they appear to be describing "two categories of pathogens." In 1985, Paul Unschuld (1985, 7) asserted that the book describes a disease called "li" that had the same symptoms as those of Hansen's disease, but he appears to modify his position somewhat when he later observes that the description of *li* "may be identified in hindsight as leprosy" (Unschuld 2003, 126). We can conclude that it is quite possible that Hansen's disease was present in China as early as the third century BCE and could have been the source for the leprosy that appeared later elsewhere in the IOW. However, the identification is uncertain.

The presence of leprosy in ancient India, as in China, is ambiguous. It is widely accepted among scholars that Hansen's disease is described in ancient Indian texts, but there is debate about which texts and how they should be dated—a debate critical to the argument that leprosy originated in India from where it spread throughout the IOW (an issue discussed below).

The first work that may refer to leprosy is the *Atharva Veda*, one of the foundational texts of Vedic Hinduism that dates to sometime between the late second and the early first millennium BCE. A.K. Sinha, B.G. Banerjee, S. Singh (2010, 1) and Gwen Robbins Schug (2016, 4) argue that the book refers to leprosy because it contains the Sanskrit word *kushtha* which is used in the later *Sushruta Samhita* to designate leprosy. This would establish the presence of leprosy in India at an extremely early date. The *Atharva Veda*, however, lacks a clinical description of *kushtha*, so it is difficult to know what skin disease or diseases are indicated (Dharmendra 1967, 2).

Numerous authors instead focus on the *Sushruta Samhita*, one of the early texts of Ayurvedic medicine. The book includes a description of *kushtha* that a minority of scholars, such as I.A. Menon and H.F. Haberman (1969, 391) and Ranes Chakravorty (1993, 410), argue covers several types of skin diseases, but which the majority believe refers to true leprosy (see, e.g. Lowe 1947, 55; Dharmendra 1967, 2). For most scholars, therefore, the debate is primarily about how to date *Sushruta Samhita*. The text is attributed to the sage Sushruta but has gone through a number of revisions, and dating the recensions has proven to be difficult. Kaviraj Bhishagranta (1907, iv) states that while there is no direct evidence for the biography of Sushruta, Nagarjuna, the editor of the current redaction of the work, is known to have lived in the fourth century BCE and argues that the original version had to have been written some two centuries earlier in order to qualify as a venerable text and justify the creation of a recension. Lowe (1947, 55), Dharmendra (1967, 1), Olaf Skinsnes (1973, 221) and Stewart Cole and Pushpendra Singh (2012, 4) accept the argument that the text was compiled around 600 BCE from older materials dating to sometime between 1000 and 800 BCE. However, Gerald Larson (1987, 247), Chakravorty (1993, 408) and Steven Engler (2003, 420) counter that the original text dates to about 600 BCE and that the final recension was completed as late as the tenth century CE. It should be noted that when briefly surveying the early history of leprosy, many authors note the possible presence of leprosy in ancient India and refer to ancient Indian texts in general, or the *Sushruta Samhita* specifically, as dating to 600 BCE, without going into the details (Browne 1975, 485; Roberts and Manchester 2005, 200; Gelber 2015). The implication of these differing interpretations is that even if we accept that Hansen's disease can be identified in the *Sushruta Samhita*, the date of the earliest literary evidence for the presence of leprosy in India could conceivably be any time between 1000 BCE and 1000 CE.

The question of the presence of leprosy in ancient Egypt, and whether Egypt is the source of leprosy within the IOW, is more contentious than for China or India. In an analysis of the Ebers papyrus (c. 1500 BCE), B. Ebbell (1935, 259) argues that the description of an affliction called Chons' swelling "agrees perfectly with modular leprosy." While not as certain as Ebbell, Skinsnes (1973, 220) believes that the Ebers papyrus may very well describe true leprosy because no one has yet proven the text did not refer to leprosy. Similarly, Michael Lechat (1999, 462) asserts that leprosy probably did exist in Egypt and places to the east at the time of the Ebers papyrus because there is no conclusive evidence to the contrary. Helen Donoghue et al. (2015b, 251) cite Lechat (1999) in support of their statement that ancient written accounts "suggest" that leprosy existed in ancient Egypt.

Also, while Cole and Singh (2012, 2) and Robbins et al. (2009, 1) cite E.V. Hulse (1972) in support of their claim that leprosy is mentioned in ancient Egyptian texts, Hulse (1975, 88) believes the opposite and argues

> as the Ebers Papyrus is essentially a prescription book, clinical details are relatively few. However, those given for the two conditions which have been suspect (Uchedu and Chons' swelling) are sufficient to rule out the possibility that they were the disease now known as leprosy.

Numerous authors including Lowe (1947, 59), Vilhelm Møller-Christensen (1967, 304), Stanley Browne (1975, 485) and Michael Dols (1979, 314) agree that there is insufficient literary evidence to establish the presence of leprosy in ancient Egypt. Johannes Andersen (1969, 11, 13) goes further in stating that Ebbell's claim that Chons' disease is leprosy is "a far fetched interpretation" demonstrating that "Ebbell does not understand leprosy."

Somewhat surprisingly, in recent years several scholars, such as Robbins et al. (2009), Cole and Singh (2012) and Donoghue et al. (2015a), have returned to the possibility that Hansen's disease may have existed in ancient Egypt. In addition, since 2003, the World Health Organization has issued a fact sheet on leprosy on its website that states "Leprosy was recognized in the ancient civilizations of China, Egypt and India" (World Health Organization 2003; see also World Health Organization 2016). The WHO fact sheet was cited by Donoghue et al. (2015a, S142) in support of their claim that "The disease has been described in ancient China, Egypt and India" but with the caveat that "it is sometimes difficult to distinguish between leprosy and other diseases with similar external symptoms" and is also probably the source of a BBC online story about excavations at the mediaeval leprosy hospital of St. Mary Magdalen, Winchester, UK, when it observed that leprosy "was recorded an ancient China, Egypt and India" (Briggs 2017). However, the WHO does not cite a source for their claim.

There is similarly scholarly disagreement as to the presence of leprosy in the early Mediterranean. As Dols (1979, 318) and David Stannard (1993, 262) point out, medical terms used in classical texts cover a far wider range of meanings than they do in modern works. The English word leprosy is derived via Latin from the Greek word *lepra* which, while used by various writers from Hippocrates to nearly the present, has not always signified Hansen's disease (Demaitre 2007, 85–102).

One persistent issue has been dating the arrival of true leprosy in the Mediterranean, and possibly determining its origin, by identifying the first author to reliably describe it. Several scholars have argued that Greek writers Aristotle (384–322 BCE), Herodotus (c. 484–430/420 BCE) and Hippocrates (c. 460–c. 377 BCE) referred in their works to leprosy. However, Herodotus (who used the word *lepra*) and Aristotle (*satyriasis*) did not describe the affliction, so it is impossible to know the disease they intended to indicate (Dols 1979, 314). Hippocrates, or more accurately the Hippocratic canon, mentions an ailment called *lepra*, but from the description given, it appears to be a collective term for several skin complaints rather than a specific designation for Hansen's disease (Andersen 1969, 17–18; Carmichael 1993, 838; Nutton 2004, 29). The first author to give a description of an affliction resembling Hansen's disease is the Roman, Celsus (Aulus Cornelius Celsus, first century CE), who used the Greek word *elephantiasis* (Elephant Disease), a usage that later authors such as fellow Romans, Pliny the Elder (Gaius Plinius Secundus, 23–79 CE) and Aretaeus of Cappadocia (second century CE), appear to have adopted (Andersen 1969, 19–30; Dols 1979, 315; Nutton 2004, 30).

The famous Roman physician, Galen (129–199 CE), used both *lepra* and *elephantiasis* in his writings, but in ways that have caused later scholars some confusion. His descriptions of the two disorders overlap, and on occasion he used both terms in the same sentence (Demaitre 2007, 16; Dols 1979, 318). The Catalan physician, Arnau de Vilanova (1235–1312), in his commentary on Galen's *De Malicia Complexionis Diverse*, recognised the ambiguity in the text and observed that Galen used *lepra* as the name of both a disease and a symptom of disease (Demaitre 2007, 117–118). Consequently, from the fourteenth century onward, European physicians turned to the *Canon of Medicine* (*Qanun*) by Persian physician, Avicenna (Ibn Sina, 980–1037), for clarification (Carmichael 1993, 837–838). In his extensive coverage of the symptoms and treatment of leprosy, Avicenna used the word *al-judam*, which was the Arabic equivalent to *elephantiasis*, but in the Latin translation, *al-judam* was rendered as *lepra* (Demaitre 2007, 88).

In recent years, several authors (Arrizabalaga 2002, Cunningham 2002, Stein 2014) have questioned the reliability and even the usefulness of combing ancient texts for anything that may resemble the symptoms of Hansen's disease. Sometimes called "retrospective diagnosis," the effort to identify diseases in the past is complicated by both the question of translation of past vocabularies of disease and differing conceptualisations of health and disease in past societies. Such factors may result in the imposition of modern understanding of disease on historic texts and thus misreading them. Andrew Cunningham (2002, 14) argues that disease is both a biological and a social phenomenon so that "we can only think about our experience of disease—as of anything else—in the terms and categories of whichever particular society we are in." As such, retrospective diagnosis may lead to a "biologist reductionism" that overlooks the social phenomena that are inherent in any study of disease (Arrizabalaga 2002, 56).

Critics of the "old and worn methodology" of retrospective diagnosis believe that it is not only outdated but also self-serving (Stein 2014, 54). Arrizabalaga (2002, 67) argues that too many scholars have engaged in overly creative interpretations of disease descriptions found in ancient texts in order to support their own theories regarding the appropriate classification of diseases. Mukharji (2014, 67) goes further in asserting that efforts at retrospective diagnosis actually seek to "retrofit" the past, and raises the question "who gets to retrofit whose past?"

In addition, critics of retrospective diagnosis argue that it is almost impossible to map modern onto ancient disease categories. As we cannot use contemporary criteria to identify ancient diseases, we should instead focus on how ancient societies conceptualised disease and how they sought to understand individual disorders (Leung 2009, 17). We would thereby have a "more historically sensitive and critical approach to distinctive forms of temporality" (Mukharji 2014, 68).

While acknowledging the validity of several criticisms of retroactive diagnosis, some authors contend that, in certain circumstances, retroactive diagnosis can be a viable and reliable methodology. Piers Mitchell (2011, 86) argues that it is possible to examine both the social and biological aspects of historical diseases because the two are inherently connected. The social aspect of disease cannot exist without some physiological process being involved.

Archaeology and the Early History of Leprosy: The Osteological Evidence

As noted above, the literary evidence for the early history of leprosy is scanty and subject to widely diverging interpretations. In an effort to find new types of evidence, scholars have turned to the analysis of the physical remains of people with leprosy. As leprosy is one of the few diseases that can cause pathological bone changes, their presence indicates that the individual was infected with *M. leprae*, that is, true leprosy. Some bone changes, such as rhinomaxillary syndrome which involves dental abscessing of the maxilla bones, ulceration and reduction of the nasal bone and nasal spine and widening of the nasal aperture, are caused directly by *M. leprae*. Others, especially the bones and joints of the feet and hands, are due to secondary infections (Taylor et al. 2013, 2). However, because changes in the hands and feet can be caused by more than one pathogen, leprosy can only be conclusively diagnosed when indisputable rhinomaxillary changes are present (Møller-Christensen 1967, 295 and Andersen et al. 1994, 21).

The diagnosis of leprosy in skeletal remains is complicated by two major factors. First, changes in the bones in the rhinomaxillary area can also be caused by treponemal disease, and to the hands and feet by psoriatic and septic arthritis and other joint diseases (Donoghue et al. 2015b, 251). Second, only roughly 5 per cent of people infected with *M. leprae* show the typical pathological bone changes, meaning that many individuals who did have leprosy cannot be identified visually (Roberts and Manchester 2005, 195). The skeletal response to leprosy varies in individuals because the disease alternates between acute and less active phases (Ortner 2002, 74). This means that differential diagnosis depends on how long the individual in question had the disease, and that there is a chance that recent infections of leprosy may not leave any trace (Møller-Christensen 1967, 300).

The vast majority of skeletons with bone damage attributable to leprosy have been found in Europe, particularly in Denmark and Britain, and date to the mediaeval era (Andersen 1969; Mendum et al. 2014). This has limited the value of archaeological discoveries in Europe for the earlier history of leprosy, as the presence of leprosy there in mediaeval times has been relatively well established through textual sources. Over the past few years, some potentially earlier specimens have been uncovered, in Britain (fifth to sixth centuries CE), Egypt (second century BCE), India and Pakistan (c. 2000 BCE), Israel (first century CE), Italy (second century CE) and Uzbekistan (first to fourth centuries CE). The Egyptian specimens were found in a cemetery located in the Dakhleh Oasis, in the southwest of the country, some 350 km from the Nile. They included four skulls with rhinomaxillary damage as well as foot bones that show evidence of damage from leprosy. However, it is unclear if the foot bones are from the same individuals as the skulls (Dzierzykray-Rogalski 1980, 73). Subsequent authors have generally accepted the diagnosis of Hansen's disease (Roberts and Manchester 2005, 201). Of the Indian specimens, one skull from Balathal showed rhinomaxillary damage, as did the nine from Harappa (Robbins et al. 2009, 1 and Robbins Schug et al. 2013, 6), and most scholars have accepted the diagnosis of leprosy (Schuenemann et al. 2013, 183; Lovell 2016, 172). The challenge now is to contextualise these findings within the broader early history of leprosy.

The Impact of Molecular Biology

According to Cheryl Andam et al. (2016, 980), the analysis of ancient DNA (aDNA) can make useful contributions to an understanding of the ancient history of disease. Moreover, Kelly Harkins and Anne Stone (2015, 137) argue that rapidly improving DNA sequencing technologies have transformed our understanding of the biological and social processes of ancient diseases. However, it is important to be aware of the distinct limitations of the use of aDNA evidence. Jones and Nevell (2016, e238) argue that scientific evidence "should... be historicised just as rigorously as other forms of historical evidence," while Andam et al. (2016, 988) caution that doubt still exists about how well genetic data can be reconciled with the existing literary and archaeological evidence. Mitchell (2011, 88) admits that most of the disorders described in ancient texts will never be identified with known diseases, but argues that there are some instances where a modern biological diagnosis may be determined. Paul Unschuld (1985, 5) asserts that this is possible with leprosy because Hansen's disease is a "real nosological unit" in the "sense that it is caused by a microorganism (*Mycobacterium leprae*) that appears to have affected patients in different cultures and continents for over two millennia." It may be possible to use this biological reality to identify evidence for the early history of leprosy.

Methods for DNA Analysis

With the uncertainties and ambiguities inherent in the interpretation of the literary and osteological evidence, scholars seeking to elucidate the early history of leprosy have turned to studying the biological reality of *M. leprae* for usable evidence. While some studies have looked at cell wall lipid biomarkers, the major focus has been on analysis of the pathogen's DNA. Numerous DNA analytical methods have been developed during the last few decades, but the ones utilised most commonly in the study of *M. leprae* are polymerase chain reaction (PCR), developed in the late 1980s, and next-generation or high-throughput approaches, first used in 2005. With PCR the analysis focuses on a single genetic loci, such as changes in the nucleotides in specific regions of the DNA. DNA is composed of four nucleotides, adenine, guanine, thymine and cytosine, and a change in one of them comprises a single nucleotide polymorphism (SNP) that can be compared with other SNPs and used to determine the species of a given pathogen. With next-generation approaches, the focus moves from a single genetic loci to genome-wide data which can be processed much faster and more reliably and can be used for genotyping or identifying different strains of the bacterium (Anastasiou and Mitchell 2013, 27–28; Andam et al. 2016, 980–982).

The use of PCR and next-generation approaches to analyse DNA extracted from the skeletal remains of persons with leprosy recovered from archaeological sites holds great promise for finding evidence relevant to the early history of leprosy. Initially, attempts were made to simply detect the presence of *M. leprae* in bone lesions. A tentative diagnosis of leprosy was then made. Some scholars, such as Michel Drancourt and Didier Raoult (2005, 23), considered such methods much more conclusive than hypotheses derived solely from archaeological or historical sources. Of course, the mere presence of *M. leprae* does not necessarily establish the cause of death of an individual but "its presence in ancient human remains provides clear evidence of infection" (Donoghue et al. 2015b, 251). The application of these methods to the analysis of aDNA, however, can be difficult in that the "DNA molecules preserved within archaeological or historic [human] remains … are fragmented, damaged, and with rare exception, dominated by 99% or more of contaminating microbial DNA sequences" (Harkins and Stone 2015, 137). In the rush to use the new and rapidly advancing technology, some of the early studies (critics like Wilbur and Stone [2012, 704] would say "most") were plagued by inadequate controls to prevent contamination of samples in the field and the laboratory, as well as weak technical standards that prevented independent verification of the data from being made (Willerslev and Cooper 2006, 643; Stone et al. 2009, 81; Anastasiou and Mitchell 2013, 28). There has also been the hitherto insoluble problem that the skeletal material available for aDNA extraction is very limited, so that the analysis of aDNA may never produce as much evidence for the early history of leprosy as was first hoped (Wilbur and Stone 2012, 711).

The Findings of DNA Analysis

The effort to recover *M. leprae* aDNA from specimens around the world has thus far produced mixed results. In several cases, aDNA has been recovered from suspected skeletal material, which confirms the presence of *M. leprae*. However, few specimens have aDNA sufficiently well preserved to be genotyped (Donoghue et al. 2015b, 251). In other cases, the skeleton shows the osteological changes associated with leprosy, but aDNA can either not be recovered (Rubini et al. 2012, 578) or is too degraded for positive genomic analysis (Lovell 2016, 173).

Studies on the genotyping of *M. leprae* first appeared in 2005. Marc Monot sequenced modern DNA from different parts of the world (Monot et al. 2005, 1040 and 2009, 1287), refined the methodology and sequenced both modern DNA and aDNA from Croatia, Denmark, Egypt, England, Hungary and Turkey. These studies found four phylogenetic groups, and their authors proposed a model in which the progenitor strain of *M. leprae* may have originated in East Africa, then accompanied humans as they migrated into Asia where, over time, new strains emerged (Monet et al. 2009, 9). This model does not necessarily conflict with the widely accepted literary and osteological evidence for early leprosy, but the phylogeny of *M. leprae* is incompatible with the hypotheses that merchants, slave girls or Alexander the Great's soldiers carried leprosy by sea from India to the Mediterranean world (Cole and Singh 2012, 9).

More recent studies have challenged this model by identifying a fifth phylogenetic group and proposing divergence times for the most recent common ancestor (MRCA) for all *M. leprae* strains of 1350 to 5078 or 1975 to 4562 years ago (Schuenemann et al. 2013, 182), 1400 to 2700 years ago (Mendum et al. 2014, 2) or around 2500 years ago (Inskip et al. 2015, 16). All of these dates are significantly more recent than those proposed by Monot et al. (2009). However, none of these studies have suggested a model for the origin and transmission of leprosy. Moreover, their findings are open to any number of interpretations. As a result, scholars are currently attempting to reconcile the new DNA evidence with existing literary and osteological evidence (Robbins Schug 2016, 1–2).

Conclusion

As we have seen, the literary and archaeological evidence for the early history of leprosy is fragmentary, ambiguous and sometimes contradictory. Many sometimes quite imaginative hypotheses have been developed to explain the emergence and spread of the disease within the IOW. Meanwhile, the latest advances in molecular biology, despite their great potential, have yet to provide the DNA evidence that can fill the gaps in the literary and archaeological record and mediate between the conflicting interpretations. Unschuld's proposition that Hansen's disease is a real nosological unit seems to have been validated by the recovery of *M. leprae* DNA from human remains dating from as early as the first century CE in Israel (Donoghue et al. 2015b, 251), first to fourth centuries CE in Uzbekistan (Taylor et al. 2009, 2414) and fifth to sixth centuries CE in Britain (Inskip et al. 2015, 1). However, these findings are few, widely dispersed in space and time, and are open to many different interpretations, and it is difficult to know how to conceptualise the current lack of aDNA samples from India and other regions in the IOW. It is clear that there are still many unanswered questions about the early history of leprosy and the role of the IOW in that history.

References

Anastasiou, Evilena, and Piers D. Mitchell. 2013. Evolutionary Anthropology and Genes: Investigating the Genetics of Human Evolution from Excavated Skeletal Remains. *Gene* 528: 27–32.

Andam, Cheryl P., Colin J. Worby, Qiuzhi Chang, and Michael G. Campana. 2016. Microbial Genomics of Ancient Plagues and Outbreaks. *Trends in Microbiology* 24: 978–989.

Andersen, Johannes G. 1969. *Studies in the Medieval Diagnosis of Leprosy in Denmark: An Osteoarchaeological, Historical, and Clinical Study*. Copenhagen: Costers Bogtrykkeri.

Andersen, Johannes G., Keith Manchester, and Charlotte Roberts. 1994. Septic Bone Changes in Leprosy: A Clinical, Radiological and Paleopathological Review. *International Journal of Osteoarchaeology* 4: 21–30.

Arrizabalaga, Jon. 2002. Problematizing Retrospective Diagnosis in the History of Disease. *Asclepio* 54: 51–70.

Beckett, D.W. 1987. The Striking Hand of God: Leprosy in History. *New Zealand Medical Journal* 12: 494–497.

Bhishagranta, Kaviraj Kunja Lal, trans. 1907. *An English Translation of the Sushruta Samhita, Vol. 1 Sutrasthanam*. Calcutta: Wilkins Press.

Bodde, Derk. 1982. Forensic Medicine in Pre-Imperial China. *Journal of the American Oriental Society* 102: 1–15.

Briggs, Helen. 2017. Skeleton Offers Clues to Medieval Spread of Leprosy. *BBC News*. http://www.bbc.com/news/science-environment–38748911. Accessed 1 Feb 2017.

Britton, Warwick J. 2017. Leprosy. In *Infectious Diseases*, ed. Jonathan Cohen, William G. Powderly, and Steven M. Opal, 954–960. Elsevier.

Brooks, E. Bruce, and A. Taeko Brooks. 1998. *The Original Analects: Sayings of Confucius and His Successors*. New York: Columbia University Press.

Browne, Stanley G. 1975. Some Aspects of the History of Leprosy: The Leprosie of Yesterday. *Proceedings of the Royal Society of Medicine* 68: 485–493.

Carmichael, Ann G. 1993. Leprosy. In *The Cambridge World History of Human Disease*, ed. Kenneth F. Kiple, 834–839. Cambridge: Cambridge University Press.
Chakravorty, Ranes C. 1993. Diseases of Antiquity in South Asia. In *The Cambridge World History of Human Disease*, ed. Kenneth F. Kiple, 408–413. Cambridge: Cambridge University Press.
Cole, Stewart, and Pushpendra Singh. 2012. History and Phylogeography of Leprosy. In *Leprosy: A Practical Guide*, ed. Enrico Nunzi and Cesare Massone, 3–18. Milan: Springer–Verlag Mailand.
Cunningham, Andrew. 2002. Identifying Disease in the Past: Cutting the Gordian Knot. *Asclepio* 54: 13–34.
Demaitre, Luke E. 2007. *Leprosy in Premodern Medicine: A Malady of the Whole Body*. Baltimore: Johns Hopkins University Press.
Dharmendra. 1967. *Notes on Leprosy*. 2nd ed. New Delhi: Government of India Ministry of Health.
Dols, Michael W. 1979. Leprosy in Medieval Arabic Medicine. *Journal of the History of Medicine and Allied Sciences* 36: 314–333.
Donoghue, Helen D., Mark Spigelman, Justin O'Grady, Ildikó Szikossy, Ildikó Pap, Oona Y.-C. Lee, Houdini H.T. Wu, Gurdyal S. Besra, and David E. Minnikin. 2015a. Ancient DNA Analysis – An Established Technique in Charting the Evolution of Tuberculosis and Leprosy. *Tuberculosis* 95 (Supplement 1): S140–S144.
Donoghue, Helen D., G. Michael Taylor, Antónia Marcsik, Erika Molnár, Gyorgy Pálfi, Ildikó Pap, Maria Teschler-Nicola, et al. 2015b. A Migration–Driven Model for the Historical Spread of Leprosy in Medieval Eastern and Central Europe. *Infection, Genetics and Evolution* 31: 250–256.
Drancourt, Michel, and Didier Raoult. 2005. Palaeomicrobiology: Current Issues and Perspectives. *Nature Reviews Microbiology* 3: 23–35.
Dzierzykray-Rogalski, T. 1980. Paleopathology of the Ptolemaic Inhabitants of Dakhleh Oasis (Egypt). *Journal of Human Evolution* 9: 71–74.
Ebbell, B. 1935. A Contribution to the Earliest History of Leprosy. *International Journal of Leprosy* 3: 257–263.
Engler, Steven. 2003. 'Science' vs. 'Religion' in Classical Ayurveda. *Numen* 50: 416–463.
Gelber, Robert H. 2015. Leprosy. In *Harrison's Principles of Internal Medicine*, ed. James F. Shanahan and Kim J. Davis, 19th ed. McGraw-Hill. http://accessmedicine.mhmedical.com/book.aspx?bookid=1130.
Harkins, Kelly M., and Anne C. Stone. 2015. Ancient Pathogen Genomics: Insights into Timing and Adaptation. *Journal of Human Evolution* 79: 137–149.
Hulse, E.V. 1972. Leprosy and Ancient Egypt. *The Lancet* 2 (7785): 1024–1025.
———. 1975. The Nature of Biblical 'Leprosy' and the Use of Alternative Medical Terms in Modern Translations of the Bible. *Palestine Exploration Quarterly* 107: 87–105.

Inskip, Sarah A., G. Michael Taylor, Sonia R. Zakrzewski, Simon A. Mays, Alistair W.G. Pike, Gareth Llewellyn, Christopher M. Williams, et al. 2015. Osteological, Biomolecular and Geochemical Examination of an Early Anglo-Saxon Case of Lepromatous Leprosy. *PLOS ONE*. https://doi.org/10.1371/journal.pone.0124282.

Jones, Lori, and Richard Nevell. 2016. Plagued by Doubt and Viral Misinformation: The Need for Evidence-Based Use of Historical Disease Images. *The Lancet* 16: e235–e240.

Larson, Gerald James. 1987. Āyurveda and the Hindu Philosophical Systems. *Philosophy East and West* 37: 245–259.

Lechat, Michael F. 1999. The Paleoepidemiology of Leprosy: An Overview. *International Journal of Leprosy and Other Mycobacterial Diseases* 67: 460–470.

Leung, Angela Ki Che. 2009. *Leprosy in China: A History*. New York: Columbia University Press.

Lloyd Davies, Margaret, and T.A. Lloyd Davies. 1989. Biblical Leprosy: A Comedy of Errors. *Journal of the Royal Society of Medicine* 82: 622–623.

Lovell, Nancy C. 2016. Bioarchaeology of the Indus Valley Civilization: Biological Affinities, Paleopathology, and Chemical Analyses. In *A Companion to South Asia in the Past*, ed. Gwen Robbins Schug and Subhas R. Walimbe, 169–186. Malden, MA: Wiley Blackwell.

Lowe, John. 1947. Comments on the History of Leprosy. *Leprosy Review* 18: 54–64.

Lu, Gwei-Djen, and Joseph Needham. 1967. Records of Diseases in Ancient China. In *Diseases in Antiquity: A Survey of the Diseases, Injuries and Surgery of Early Populations*, ed. Don Brothwell and A.T. Sandison, 222–237. Springfield, IL: Charles C. Thomas.

Manchester, Keith. 1984. Tuberculosis and Leprosy in Antiquity: An Interpretation. *Medical History* 28: 162–173.

Mark, Samuel. 2002. Alexander the Great, Seafaring, and the Spread of Leprosy. *Journal of the History of Medicine and Allied Sciences* 57: 285–311.

McClish, Mark. 2012. Is the *Artaśāstra* a Mauryan Document? In *Reimagining Aśoka: Memory and History*, ed. Patrick Olivelle, Janice Leoshko, and Himanshu Prabha Ray, 280–309. Oxford: Oxford University Press.

Mendum, Tom A., Verena J. Schuenemann, Simon Roffey, G. Michael Taylor, Huihai Wu, Pushpendra Singh, Katie Tucker, et al. 2014. *Mycobacterium leprae* Genomes from a British Medieval Leprosy Hospital: Towards Understanding an Ancient Epidemic. *BMC Genomics*. https://doi.org/10.1186/1471-2164-15-270.

Menon, I.A., and H.F. Haberman. 1969. Dermatological Writings of Ancient India. *Medical History* 13: 387–392.

Miller, Timothy S., and John W. Nesbitt. 2014. *Walking Corpses: Leprosy in Byzantium and the Medieval West*. Ithaca, NY: Cornell University Press.

Mitchell, Piers D. 2011. Retrospective Diagnosis and the Use of Historical Texts for Investigating Disease in the Past. *International Journal of Paleopathology* 1: 81–88.

Møller-Christensen, Vilhelm. 1967. Evidence of Leprosy in Earlier Peoples. In *Diseases in Antiquity: A Survey of the Diseases, Injuries and Surgery of Early Populations*, ed. Don Brothwell and A.T. Sandison, 295–306. Springfield, IL: Charles C. Thomas.

Monot, Marc, Nadine Honoré, Thierry Garnier, Romulo Araoz, Jean-Yves Coppée, Céline Lacroix, Samba Sow, et al. 2005. On the Origin of Leprosy. *Science* 308: 1040–1042.

Monot, Marc, Nadine Honoré, Thierry Garnier, Nora Zidane, Diana Sherafi, Alberto Paniz-Mondolfi, Masanori Matsuoka, et al. 2009. Comparative Genomic and Phylogeographic Analysis of *Mycobacterium leprae*. *Nature Genetics* 41: 1282–1289.

Mukharji, Projit Bihari. 2014. Vishalyakarani and *Eupatorium ayapana*: Retro-botanizing, Embedded Traditions and Multiple Historicities of Plants in Colonial Bengal, 1890–1940. *The Journal of Asian Studies* 73: 65–87.

Needham, Joseph with Lu, Gwei-Djen. 2000. *Science and Civilization in China, Volume 6 Biology and Biological Technology, Part VI: Medicine*. Ed. Nathan Sivin. Cambridge: Cambridge University Press.

Nutton, Vivian. 2004. *Ancient Medicine*. New York: Routledge.

Olivelle, Patrick, ed. 2013. *King, Governance, and Law in Ancient India: Kauṭilya's Artaśāstra*. Oxford: Oxford University Press.

Ortner, Donald J. 2002. Observations on the Pathogenesis of Skeletal Disease in Leprosy. In *The Past and Present of Leprosy: Archaeological, Historical, Paleopathological and Clinical Approaches*, ed. Charlotte A. Roberts, Mary E. Lewis, and K. Manchester, 73–80. BAR International Series 1054.

Robbins Schug, Gwen. 2016. Begotten of Corruption? Bioarchaeology and 'Othering' of Leprosy in South Asia. *International Journal of Paleopathology* 15: 1–9.

Robbins Schug, Gwen, K. Elaine Blevins, Brett Cox, Kelsey Gray, and V. Mushrif-Tripathy. 2013. Infection, Disease, and Biosocial Processes at the End of the Indus Civilization. *PLOS ONE*. https://doi.org/10.1371/journal.pone.0084814.

Robbins, Gwen, V. Mushrif Tripathy, V.N. Misra, R.K. Mohanty, V.S. Shinde, Kelsey M. Gray, and Malcolm D. Schug. 2009. Ancient Skeletal Evidence for Leprosy in India (2000 B.C.). *PLOS ONE*. https://doi.org/10.1371/journal.pone.0005669.

Roberts, Charlotte, and Keith Manchester. 2005. *The Archaeology of Disease*. 3rd ed. Ithaca: Cornell University Press.

Rubini, Mauro, Yilmaz S. Erdal, Mark Spigelman, Paola Zaio, and Helen D. Donoghue. 2012. Paleopathological and Molecular Study on Two Cases of Ancient Childhood Leprosy from the Roman and Byzantine Empires. *International Journal of Osteoarchaeology* 24: 570–582.

Schuenemann, Verena J., Pushpendra Singh, Thomas A. Mendum, Ben Krause-Kyora, Günter Jäger, Kirsten I. Bos, Alexander Herbig, et al. 2013. Genome-Wide Comparison of Medieval and Modern *Mycobacterium leprae*. *Science* 341: 179–183.

Sinha, A.K., B.G. Banerjee, and S. Singh. 2010. Leprosy and Its Socio-Cultural Perception in Indian Religions and Ancient Texts. *Indian Journal of Leprosy* 82: 1–21.

Skinsnes, Olaf K. 1973. Notes from the History of Leprosy I. Interpretive Chronology of Leprosy Concept and Practice. *International Journal of Leprosy* 41: 220–233.

Stannard, David E. 1993. Disease, Human Migration, and History. In *The Cambridge World History of Human Disease*, ed. Kenneth F. Kiple, 35–42. Cambridge: Cambridge University Press.

Stein, Claudia. 2014. 'Getting' the Pox: Reflections by an Historian on How to Write the History of Early Modern Disease. *Nordic Journal of Science and Technology Studies* 2: 53–60.

Stone, Anne C., Alicia K. Wilbur, Jane E. Buikstra, and Charlotte A. Roberts. 2009. Tuberculosis and Leprosy in Perspective. *Yearbook of Physical Anthropology* 52: 66–94.

Taylor, G. Michael, Soren Blau, Simon Mays, Marc Monot, Oona Y.-C. Lee, David E. Minnikin, Gurdyal S. Besra, Stewart T. Cole, and Paul Rutland. 2009. *Mycobacterium leprae* Genotype Amplified from an Archaeological Case of Lepromatous Leprosy in Central Asia. *Journal of Archaeological Science* 36: 2408–2414.

Taylor, G. Michael, Katie Tucker, Rachel Butler, Alistair W.G. Pike, Jamie Lewis, Simon Roffey, Philip Marter, et al. 2013. Detection and Strain Typing of Ancient Mycobacterium leprae from a Medieval Leprosy Hospital. *PLOS ONE*. https://doi.org/10.1371/journal.pone.0062406.

Unschuld, Paul U. 1985. Traditional Chinese Medical Theory and Real Nosological Units: The Case of Hansen's Disease. *Medical Anthropology Quarterly* 17: 5–8.

———. 2003. *Huang Di Nei Jing Su Wen: Nature, Knowledge, Imagery in an Ancient Chinese Medical Text*. Berkeley and Los Angeles: University of California Press.

Virmond, Marcos, Andrzej Grzybowski, and Luiza Virmond. 2015. Leprosy: A Glossary. *Clinics in Dermatology* 33: 8–18.

Wilbur, Alicia K., and Anne C. Stone. 2012. Using Ancient DNA Techniques to Study Human Disease. In *The Global History of Paleopathology*, ed. Jane E. Buikstra and Charlotte A. Roberts, 703–717. Oxford: Oxford University Press.

Willerslev, Eske, and Alan Cooper. 2006. Reply. Pathogenic Microbial Ancient DNA: A Problem or an Opportunity? *Proceedings of the Royal Society B* 273: 643.

World Health Organization. 2003, January. Leprosy. Fact Sheet 101. https://web-beta.archive.org/web/20030210102100/http://who.int/mediacentre/factsheets/fs101/en. Accessed 6 Feb 2017.

———. 2016, October. Leprosy. Fact Sheet 101. http://who.int/mediacentre/factsheets/fs101/en. Accessed 16 Jan 2016.

Zimmerman, Susan. 2008. Leprosy in the Medieval Imaginary. *Journal of Medieval and Early Modern Studies* 38: 559–587.

CHAPTER 5

Climate, Weather and Pestilence in the Philippines Since the Sixteenth Century

James Francis Warren

Introduction

The struggle for existence in the Philippines since the end of the sixteenth century has been precariously waged on two fronts—against inadequate food supply and associated problems linked to distribution and colonial and capitalist institutions, and against various forms of disease (Braudel 1981, 90–91; Newson 2009). It is the latter with which this chapter is concerned. In both the colonial and post-independence periods, typhoons, floods and droughts were often followed by the outbreak of disease, the incidence of which has intensified since the start of the twentieth century due to rapid climate change, population growth and commercialisation of agriculture. A rise in extreme typhoon events, warmer atmospheric temperatures and more variable rainfall patterns associated with the El Niño Southern Oscillation have increased the health consequences and local–regional risks of disease, as well as of deaths due to flood and drought. The

J. F. Warren (✉)
Murdoch University, Perth, WA, Australia
e-mail: J.Warren@murdoch.edu.au

G. Campbell, E.-M. Knoll (eds.), *Disease Dispersion and Impact in the Indian Ocean World*, Palgrave Series in Indian Ocean World Studies, https://doi.org/10.1007/978-3-030-36264-5_5

health impacts of recent extreme weather have been particularly severe for segments of the Philippine population located in areas vulnerable to typhoons and floods.

The onset of the rainy season traditionally brought flooding and health problems ranging from coughs, fever and flu to water-borne diseases such as cholera, typhoid fever, amoebiosis and diarrhoea. For example, during the damp conditions of the rainy season the common cold afflicted most Filipinos. An article entitled 'The Department of Health Warning and the Rainy Season' (*Manila Bulletin*, 18 May 2000) asserted that in colonial times, there were no proven cures for the illnesses that broke out in the aftermath of heavy rains, except proper rest and nutrition. The onset of the rains also heralded the beginning of the typhoon season, as well as water-borne and mosquito-borne diseases which broke out simultaneously. For instance, the *Manila Bulletin* of 22 July 2002 noted that steady rains and floods, especially in the lowlands, created fertile pools and breeding sites for the *Aedes* mosquito and dengue. Mortality due to cholera, diarrhoea, dysentery, typhoid, El Tor and pneumonia usually rose markedly during the typhoon season. In addition, the incidence of long dry spells during El Niño events, sometimes lasting years, even decades, caused widespread outbreaks of gastro-intestinal diseases due to the drying up of rivers and wells, which forced agricultural and tribal communities to drink unsafe water.

The Weather, Colonialism and Disease

In highlighting links between climate, weather and pestilence in the Philippines, I draw for inspiration on the path-breaking works of Alfred W. Crosby. He contends that European colonisers were successful in most temperate zones of the world—North America, Australia and New Zealand—because they dispersed demographically and spread their pathogens, pests, domesticated animals and weeds (Crosby 1972, 1986). As Thomas Griffiths and Libby Robin note, Crosby underscores the 'passive or distracted role of humans in ecosystems,' an approach which downplays the deliberate actions of humans and focuses on the 'independent and semi-independent dynamism of the natural world, itself normally the passive background in historical narratives' (Griffiths and Robin 1997, 2).

Crosby (1972) suggests that although relatively isolated indigenous peoples in the non-Western world possessed the genetic traits to acquire immunity to Old World infections, for the first 150 years of contact with

Europeans, the cycles of storm-, flood- and drought-induced illnesses came in waves, with too much diversity for Filippinos to develop immunity to them. Indeed, there is little evidence to suggest that in the Philippines the overall level of intestinal, diarrhoeal, cholera, typhoid, dengue, influenza, pneumonia and acute infectious diseases—due to the variability of climate and the weather—declined before the advent of the twentieth century.

Typhoons, Floods and Diseases

According to historians such as Luis Dery (2006) and Ken de Bevoise (1995), the history of Filipinos dwelling along rivers is replete with epidemics. For example, Cagayan, Ilocos, Pampanga, Albay and Samar experienced an unending struggle with typhoon- and flood- induced epidemics triggered by torrential rains and overflowing rivers. Map 5.1 gives an idea of the areas susceptible to typhoons, floods and storm surges. It was com-

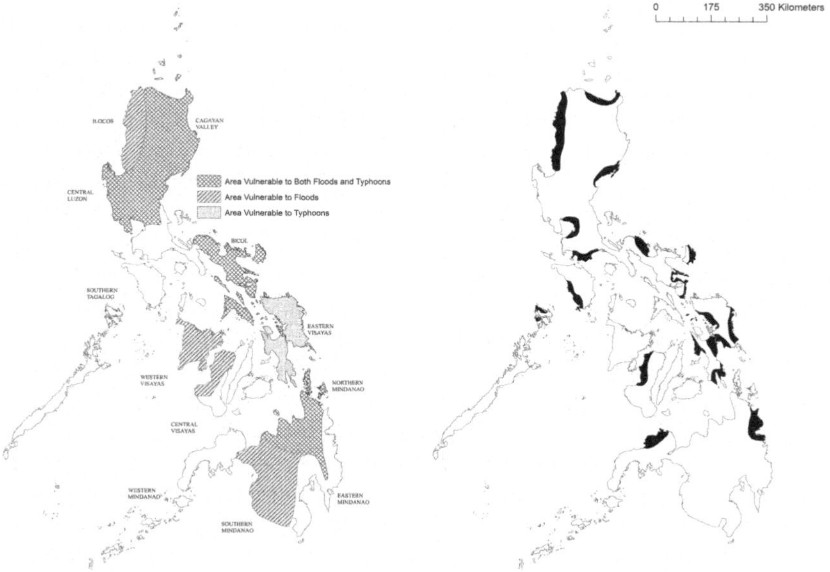

Regions vulnerable to floods and typhoons   Areas (dark) historically hit by storm surges

Map 5.1 Regions in the Philippines historically hit by floods, typhoons and storm surges. (Source: Maps created by © Julian Tyne, March 2018)

mon in these areas to find respiratory and water-borne diseases such as pneumonia, cholera, diarrhoea, dysentery and typhoid, which rose in the typhoon season, as rivers and towns flooded. Records from the sixteenth century demonstrate that typhoons and floods left hundreds, sometimes thousands, homeless. Over recent decades, millions have been affected by weather systems—as demonstrated in late 2013 by Super Typhoon Haiyan which killed over 6000 people and affected nearly 13 million.

The originally small and dispersed population of the Philippines meant that during the early colonial period, the impact of epidemics was limited. According to Linda Newson (1998), it was not until the nineteenth century that, due to the expansion of commercialised agriculture, the Philippines experienced rapid population growth and improved transport and communications, and epidemics and crisis mortality increased dramatically. Indeed, at the end of the century, *laissez faire*, market-driven colonial economic expansion 'collapsed into a decade of virulent epidemics, mass mortality and the destruction of the country's plow animals' (Doeppers and Xenos 1998, 13).

Typhoon-induced disease certainly struck more often among the poor in low-lying locales alongside rivers where flooding could take weeks to subside. Where possible, the rich lived in privileged higher-altitude locations, but they were not always spared. In some locations there was either no higher ground or the flood levels would intensify. Bridges were often destroyed and roads were either washed away or submerged, and thus it was not always possible to flee from spreading diseases, mosquito-borne viruses and starvation in the wake of a storm.

The painful task of rapidly burying hundreds of bodies—often of families and friends—and rebuilding homes destroyed by typhoons and floods often ceased at the first sign of an outbreak of disease. As fear of contagion gripped particular towns and regions, unburied corpses—human and animal—contaminated the storm-ravaged landscape.[1] Bodies lay along the roads and footpaths and infection spread because of contaminated drinking water and disease vectors, such as mosquitos, that thrived in wet conditions and swarmed over unburied corpses—especially in isolated places with little hope of emergency disaster relief. A major focus of rescue and relief teams, when or if they arrived, was to isolate

[1] A chorela epidemic hit Manila between August and October 1882, infecting 15,000 to 20,000 victims and causing widespread panic (Worcester 1908, 10).

contaminated individuals, place those suspected of carrying disease in quarantine, and remove disease-ridden corpses.[2]

In his article 'Typhoons and "Coping with Tyranny of the Urgent,"' Juan J. Mercado noted that up until the 1850s, local communities were not directly involved in taking systematic public health measures to reduce the impact of repeated disasters, especially epidemics that erupted in the aftermath of a typhoon or flood (*Today* 12 November 2001, 9). The ravages of diseases like dysentery, acute diarrhoea, influenza, typhoid fever, dengue and smallpox, plus the impoverishment caused by floods and drought, meant that successive disasters greatly accentuated the difficulty of recovery.

William McNeil, in *Plagues and People* (1979), argued that diseases and their minute organisms—germs, viruses and bacilli—appear and establish themselves on a cyclical basis, depending on the degree to which a population acquires natural defences against them (see also Braudel 1981). It is difficult to collate reliable figures for the loss of life from infectious disease in the early Spanish period, but it is important to note that more people died from infectious diseases than as a direct consequence of either a typhoon or flood. When the diseases struck and large numbers of people perished, those surviving fled, carrying the microbes to other defenceless communities.

Smallpox

A major killer was smallpox, an acute contagious viral disease, spread by airborne germs which are dispersed through sneezing and coughing. It was probably introduced into the Phillipines in the sixteenth century, if not earlier, by merchants or pirates from China or Japan, but was first identified by the Spanish in 1574 when an epidemic, called by officials a *grand enfermedad* or 'great sickness,' and by locals *bolotong*, erupted and rapidly spread from the province of Tondo to Pampanga and Pangasinan (Newson 2009, 27). The disease caused high mortality, often in epidemic proportions, affecting a number of communities, sometimes an entire region. Infection was promoted by overcrowding, enhanced travel and communication, and by dirty and dust-laden air, especially common in times of drought. All three conditions were present in the seventeenth century.

[2] For a recent case see Gladstone's (2013) *New York Times* article describing the aftermath of Haiyan, where the almost-eradicated diseases were resurfacing, along with high death rates, poor sanitation and water supply. On the lingering aftermath of Typhoon Haiyan also see Warren (2015).

Luis Dery (2006) indicates that smallpox was rampant in central Luzon between 1651 and 1668, and the disease spread rapidly as a result of trade, warfare and coerced labour. It was particularly marked around Manila and in central Luzon, where it decimated communities inhabitating the narrow coastal plain and riverine lowlands. Outbreaks occurred during prolonged periods of conscripted shipbuilding (Dery 2006, 92), and the disease was transferred long distances on the Manila galleons linking Manila and Acapulco in Mexico.

In the first half of the eighteenth century, two smallpox epidemics broke out in Tayabas, during an El Niño event, in 1715 and in 1740. Epidemics erupted again in 1757 on the central flood plain of Pampanga, in Batangas in 1774, and in 1786–1787, 1791 and 1796 (all during El Niño events) in Tayabas where it caused high mortality. Further outbreaks occurred in the Ilocos region in 1788 and in 1789 in Pampanga and Batangas (see Map 5.2).

In the early colonial period, the containment of a smallpox outbreak was virtually impossible in urban centres because of overcrowding, unrestricted inter-island travel and kinship rules which prohibited rejection of family members even if sick. Quarantine was rarely an option. Also, fear of smallpox occasionally caused Filipinos to leave their dead unburied. At such times the decomposing cadavers and germ-filled air led to further outbreaks.

Between 1574 and 1796, there was a marked decline of certain populations (see Table 5.1, and Newson 1998, 25–27) principally on Luzon, attributed to 21 smallpox epidemics.[3] However, 15 epidemics of unknown illnesses appear to have had a most disastrous effect between 1656 and 1661. These epidemics of unidentified diseases, at least two of which (1656 and 1661) occurred in El Niño years, provoked a *gran mortalidad* astronomical death rate in Ilocos, Pangasinan, Pampanga, Laguna and Tayabas. The term *gran mortalidad* was also used to describe the large-scale epidemics of 1654 and 1656 to 1660 that had disastrous effects on the populations of not only Ilocos, but Pangasinan, Pampanga, Laguna and Tayabas. Four comparable outbreaks occurred during the eighteenth century, again all during El Niño events, in 1701 and 1703 in Tondo, Pampanga and Pangasinan, and two others in Cagayan in 1767 and 1791 (see Map 5.3).

[3] See Wachtel (1977) for a comparative perspective on the Spanish colonialisation of Peru and the trauma and impact of diseases that led to biological and social collapse in the period between 1530 and 1580.

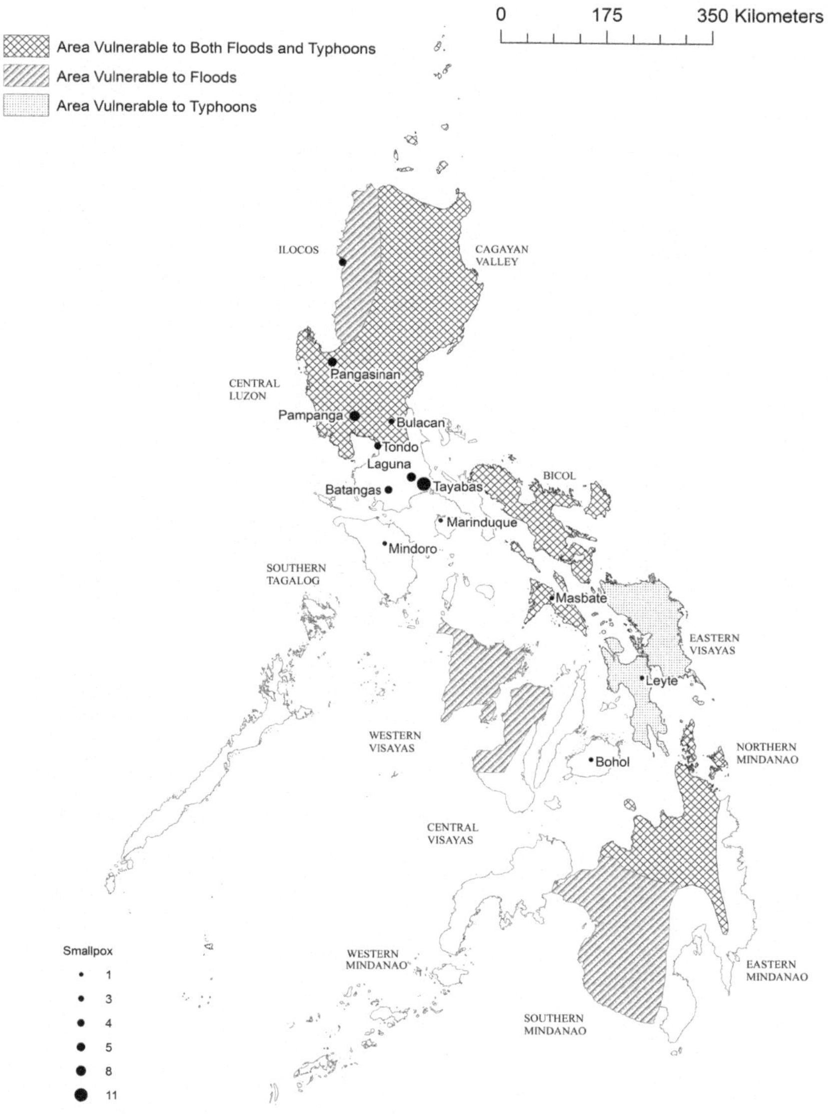

Map 5.2 Outbreaks of smallpox in the Philippines, 1574–1796. (Source: Map created by © Julian Tyne, March 2018)

Table 5.1 Smallpox and pestilence 1574–1796

Year	*Disease/climate event*	*Places affected*	*El Niño*
1574	Smallpox	Tondo, Pampanga, Pangasinan	1574
1651–1652	Smallpox	Laguna	1650
1652	Smallpox—drought	Tondo, Pampanga	1652
1653	Smallpox—drought	Ilocos, Pampanga, Tayabas	
1654	*Gran mortalidad* Drought, floods	Ilocos, Pangasinan, Pampanga	
1656	*Gran mortalidad*	Ilocos	1655
1657	Smallpox *gran mortalidad*	Ilocos, Tondo, Laguna, Mindoro, Masbate, Marinduque, Leyte, Bohol	
1658	*Gran mortalidad*	Laguna, Ilocos	
1659	*Gran mortalidad*	Pangasinan, Tayabas	
1660	*Gran mortalidad*—floods	Ilocos, Pampanga, Tayabas	
1660–1661	*Gran mortalidad*	Ilocos, Zambales, Pangasinan, Pampanga, Tayabas	1661
1664	Smallpox	Laguna, Batangas, Tayabas	
1666	Smallpox	Tayabas	
1668	Smallpox	Laguna, Batangas, Tayabas	
1679	*Gran mortalidad*	Pampanga, Tayabas	
1683	*Gran mortalidad*	Tondo	1683–1684
1691	*Gran mortalidad*, floods	Ilocos, Pampanga, Tondo	
1691–1695	*Gran mortalidad*	Ilocos	1692, 1694–1695
1701	*Gran mortalidad*—floods	Tondo, Laguna	1701
1703	*Gran mortalidad*—typhoons	Pangasinan, Pampanga, Tondo	1703–1704
1715	Smallpox, typhoons	Bulacan, Tayabas	1715–1716
1740	Smallpox	Tayabas	
1753–1756	Smallpox	Ilocos, Pangasinan, Pampanga, Bulacan, Laguna, Tayabas	1754–1755
1757	Smallpox—drought	Pampanga, Bulacan	
1759–1760	Smallpox, floods	Pangasinan, Pampanga	
1767	*Gran mortalidad*	Cagayan	1765–1766
1773	Smallpox—floods	Pampanga, Tondo	1772–1773
1774	Smallpox	Batangas	
1786	Smallpox	Pangasinan, Tayabas	1785–1786
1787	Smallpox	Tayabas	
1788	Smallpox	Ilocos, Pangasinan	
1789	Smallpox	Pampanga, Batangas	
1791	Smallpox	Tayabas	1790–1793
1791	*Gran mortalidad*	Cagayan	1790–1793
1796	Smallpox	Tayabas	1794–1797

Source: The table has been compiled from the following sources: Quinn and Neal 1992, 623–648; Dery 2006, 225–233

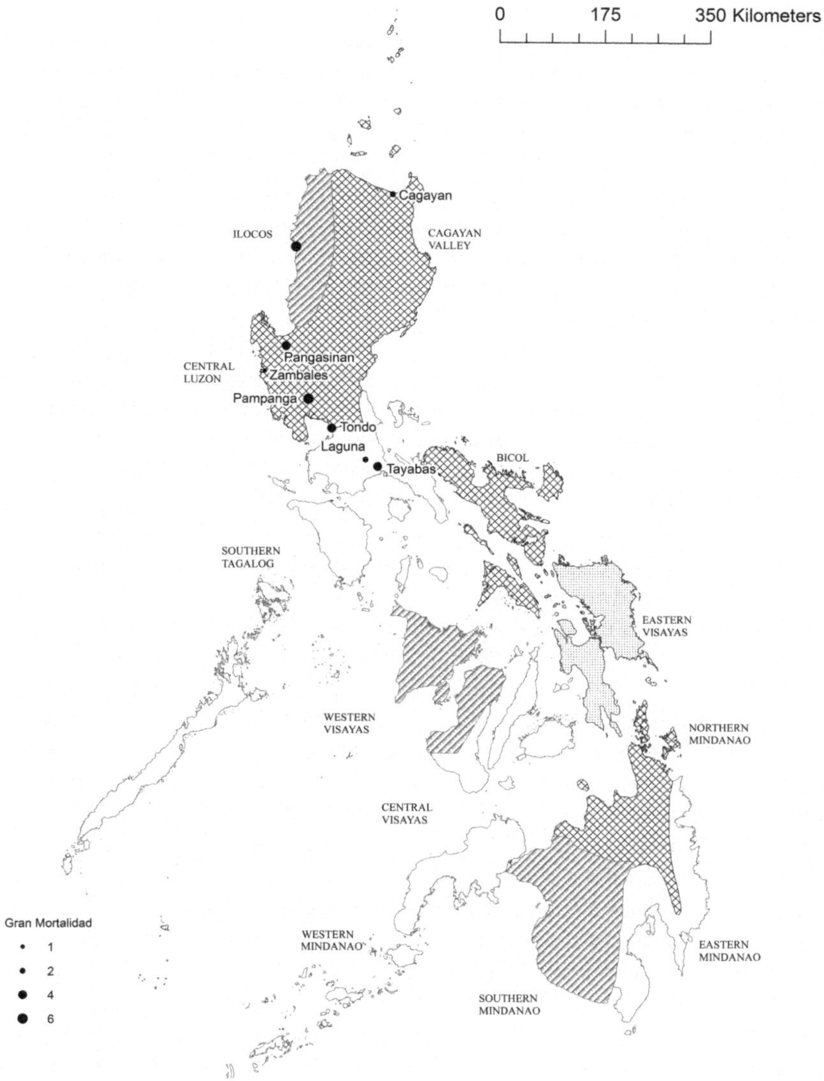

Map 5.3 Outbreaks of *gran mortalidad* in the Philippines, 1574–1796. (Source: Map created by © Julian Tyne, March 2018)

Equally damaging was the Spanish world view that any disease outbreak—smallpox, cholera or other contagious diseases—was an act of divine retribution for lack of piety in the lives of those stricken with illness. A common method of combating pestilence was to parade the image of a saint through a stricken village, in order to bring the healing of God to the beleaguered community. However, this spiritual practice, no matter how well intentioned, inadvertently contributed to the spread of the disease because the stricken populous were gathered together in a confined space of a church or convent, in order to pray for the epidemic to end. The resulting masses of ill and dying people in a confined space undoubtedly played a role in spreading the disease.

The Historical Data Papers, accounts of local and regional history compiled by the government in the 1950s and held at the National Library of the Philippines, give an excellent insight into the incidence and spread of smallpox epidemics towards the end of the Spanish period and the years under American rule. Smallpox was particularly rife in the Visayas, where efforts were being made between 1870 and 1940 to increase agricultural productivity and trade. During the years between 1898 and 1945 smallpox ranked only second to cholera in the number of epidemics reported in Visayan *barrios* and townships.

Smallpox was a major killer in the course of the Philippine–American war on Samar between 1900 and 1903, particularly in Oquerdo and Borongan. In the Philippines as a whole, the number of smallpox deaths rose from 14,860 in 1902 to 20,359 in 1903 (United States Bureau of Census 1905, 10). In 1918, a serious outbreak also occurred during a famine in Sulat. Again, both typhoid and smallpox epidemics and hunger/famine were reported in the Historical Data Papers for the Visayas, in 1900, 1904, 1912, 1918, 1919 and 1926, all El Niño years ('Reports of Epidemics and Hunger/Famine, 1802–1951').

Warwick Anderson's imaginative history, *Colonial Pathologies*, describes the cultural conflicts that took place in the Visayas, and elsewhere, as American doctors and scientists assisted by the armed forces attempted to convince a population of seven million Filipinos inhabiting more than seven thousand islands to take certain measures in order to control the spread of epidemic diseases like smallpox and cholera. But they met staunch resistance from wary farmers and poor townspeople on islands like Samar and Leyte, who were not prepared to be reformed and 'civilised' as 'biomedical subjects' (Anderson 2007; Worcester 1908). Soldiers accompanying vaccinators tried to track down those who actively resisted. In Batangas in 1902, for example, vaccinators would quarantine crowded houses, establishing a cordon sanitare, and vaccinate each person as they were filed through the doorway, if they did not show 'pock-marks' (Anderson 2007, 9).

Water-Borne Diseases, Storms and Floods

While there was a progressive increase in population from the late eighteenth century to the 1870s in the Philippines, there was not a comparable increase in land ownership by farmers and few protective measures to combat the destructive effects of typhoons and floods, such as the construction of drainage systems. More densely populated areas designated for monocrop agriculture affected by storms and flooding were increasingly susceptible to outbreaks of enteric diseases and respiratory infections due to extensive contamination of water. Spanish authorities implemented measures to prevent the recurrent floods in Pangasinan, Albay and the Visayas, but elsewhere up to the 1880s and 1890s, many people, sometimes entire regions, were stricken with pneumonia, influenza, gastro-enteritis, diarrhoea, diphtheria, cholera and typhoid (Dery 2006; Newson 2009).

Under American rule, medical emergency teams and sanitary engineers, armed with chlorine for water treatment, anti-diarrhoeal medicine and vaccines for immunisation against enteric diseases, attempted to eradicate cholera, typhoid, acute diarrhoea, dysentery and typhus. However, the contaminated environments of refuse, water, sewage and pollution in rapidly deteriorating rural settings posed major problems for American health officials.

As noted below (Table 5.2), 11,605 people died from dysentery from 1918 to 1922. Again, two dysentery epidemics broke out in Samar and Leyte in 1923 because of contaminated water supplies, causing many deaths among school children below the age of 15 ('Report of the Secretary of Public Instruction,' 1925, 117). In flood-prone areas in Cagayan, Albay, Leyte and Samar, local people often had no choice but to use contaminated water, despite official warnings against doing so. Indeed, this continues to be the case, as noted in '"Juan" Death Toll Rises to 9; Diarrhea Kills 10 in Metro Manila' (*Philippine Daily Inquirer*, 23 July 2002). The disease again reached epidemic proportions in 1924 and 1926 (when it coincided with an El Niño event), causing in total the deaths of 9339 people. By contrast, between 1918 and 1922 typhoid killed an average of 3198 people annually, and diphtheria a total of 344.

Table 5.2 Dysentery cases 1918–1928

Year	*Manila*		*Provinces*[a]		*Total*	
	Cases	*Deaths*	*Cases*	*Deaths*	*Cases*	*Deaths*
Average of five years, 1918–1922	–[b]	327	–[b]	11,278	_	11,605[c]
1923	224	115	–[b]	7051	–[b]	7166
1924	239	147	16,304	8366	16,543	8453
1925	195	124	8999	4801	9194	4925
1926	494	266	16,554	9073	17,048	9339
1927	422	242	15,204	6285	15,626	6527
1928[d]	240	132	9830	3863	10,133	3995

Source: 'Report of the Secretary of Public Instruction' (1930, 83–85)

[a]Includes transient residents registered in Manila

[b]Data not available

[c]Includes deaths with unknown residence

[d]Incomplete

During the 1920s and 1930s, according to the 'Report of the Governor General Philippine Islands, 1926,' the prevalence of intestinal infection suggested that the concerted public health initiatives and campaigns launched by the American colonial officials could not control water-borne diseases (1928, 6).

Despite efforts by the United States in various fields of science and tropical medicine to develop an adequate public health system, the post-war Philippines remained a nursery for some diseases after 1945, as the population increased. Filipinos continued to experience typhoon-induced infectious diseases throughout the second half of the twentieth century.

The post-war governments and provincial authorities found it difficult to take effective measures to eradicate these diseases, despite an unprecedented rise in the death rate from endemic diseases in the aftermath of extreme weather events. The Manila authorities consistently chose alternatives that minimised costs in providing low-cost housing, clean water supply and proper sanitation services. Modern sewage disposal methods and a clean water supply were still not available to the majority of Filipinos, even by the 1970s.

In the tumultous decade of the 1970s, when three tightly-spaced La Niña events occurred (1970–1971, 1973–1974, 1975–1976), typhoons and floods caused severe health and sanitation problems. Vulnerable individuals were once again forced to drink unsafe water and, as a result, some died from acute diarrhoea and other water-borne diseases. Most victims continue to be children whose parents and local officials have difficulty preventing from playing in dirty flooded streets, canals and creeks, or from drinking contaminated water (Atienza 2001, 2).

Cholera

Cholera, an acute diarrhoeal infection, is caused by faecal or oral contamination from ingesting contaminated water or food, has an incubation period of less than five days and can result in death through severe dehydration and kidney failure. Its symptoms are horrific:

> Radical dehydration meant that a victim shrank into a wizened caricature of his [*sic*] former self within a few hours, while ruptured capillaries discoloured the skin, turning it black and blue. The effect was to make mortality uniquely visible: patterns of bodily decay were exacerbated and accelerated, as in a time lapse motion picture, to remind all who saw it of death's ugly horror and inevitability. (McNeill 1979, 240–241)

Indeed, of all water-borne diseases, cholera terrified Filipinos the most.

For centuries, the disease was restricted to the Ganges Delta from where, in 1817, it erupted in a series of global pandemics (cf. Green and Jones this volume). Francisco Masip y Valls, Medical Director of the Civil Hospital in Manila, stated that cholera reached the Philipines in the first pandemic of 1817 (Worcester 1908, 8; Newson 2009, 17). If not then, it certainly reached the Philiipines in 1820 aboard the English frigate *Cleopatra*, from Madras. It spread in epidemic form, first in Manila, and then in most provinces across the archipelago, killing thousands of people between October 1820 and January 1822. Its impact was accentuated by three consecutive typhoons in November 1821, which also happened to be an El Niño year. Jesuit meteorologist Miguel Selga, in 'Los baguios de Noviembre de 1821 en Samar' (1939, 301), stated that due to the combined impact of the cholera epidemic and typhoons, 19 towns on Samar could not pay their taxes or fulfil obligations for personal services, and the inhabitants of many villages fled to the interior to avoid recrimination from the authorities. The Alcalde Mayor in Catbolongan described the events as a humanitarian crisis, and promised not to punish anyone who had fled to the mountains. He brought the emergency situation under control, persuading those dislocated people to return to their respective towns. He then cajoled the wealthier among them to pledge help with the costs to contain the cholera epidemic and for the rehabilitation of the province (ibid.). In 1820 in Manila the Dominicans, government officials and concerned citizens all desperately worked together to control the epidemic. Medicine for cholera victims, such as opium-based formulas, were prescribed free of charge by all the apothecary shops of the city ('The Dominicans and the epidemic of cholera at Manila in 1820').

Cholera remained endemic in the Philippines into the early twentieth century. It broke out during the rainy season and generally worsened as the typhoon season progressed. Spanish and later American officials feared that the faecal matter deposited in open spaces along flooded rivers would be carried into the public water supply, potentially contaminating it with the bacteria that causes cholera. This underscored the failure of the Spanish colonial administration in the Philippines to offer adequate health and sanitation services. Indeed in the second half of the eighteenth century, the colonial economy of the Philippines was on the verge of disintegration. During the nineteenth century, services deteriorated as the traditional Spanish system of mud drains proved utterly inadequate at a time of rapid demographic growth ('Report of the Secretary of the Interior' 1904, 115–117). Provincial committees invariably lacked sufficient funds to maintain major buildings used as hospitals and shelters (often schools or churches), and often makeshift hospitals had to operate under apallingly crowded conditions.[4] In addition, equipment was frequently old or non-existent, medical supplies limited and the water quality poor. When epidemics assumed a provincial or regional dimension, as in 1820, the Spanish government in Manila relied heavily on the assistance of the various religious orders and the charitable donations of the wealthy for disaster relief and assistance (Selga Papers, Box 5).

In total, the Phillipines experienced seven major cholera outbreaks in the nineteenth century, in 1820–1823, 1830, 1842, 1854, 1863–1865, 1882–1883 and 1888 (Worcester 1908, 4). Five of these coincided with El Niño events (1821, 1830, 1854–1855, 1864 and 1888–1889) and at least one extreme typhoon. The great typhoon of October 1882 that struck Manila and neighbouring provinces was preceded by a major cholera epidemic that reportedly arrived in Zamboanga aboard the merchant

[4] For examples of the Spanish lacking funds and the conditions faced by the populous, see AMO Selga Papers (n.d., 570), Dery (1997, 76–77, 83), PNA (1831, 1893, n.d.), Selga (1920, 1935, 42).

steamer and spread from Maybun on the island of Jolo in the Sulu Archipelago (Selga, 'Accidentes Atmosphericas Consiguados en los Libros Canonicos de la Iglesia y Legajos Historicas del Pueblo Tanay, Rizal' n.d.), and subsequently the steamer *Francisco Reyes* carried it to Manila (Worcester 1908, 6). The daily number of deaths from cholera rose from 22 on 20 August 1882 to a peak of 339 on 2 September, after which the epidemic declined—although there is evidence that officials failed to report true fatality figures (ibid., 11). Whereas official figures stated that the number of cholera-related deaths in Manila was between 2108 and 5413, the German physician M. Koeniger claimed a far higher number of victims:

> From August to October, 1882, Manila was visited by a severe cholera epidemic, which there found fertile soil, carrying off fifteen to twenty thousand victims. A panic occurred among the natives as well as among the European population, because cholera had not been present since 1865, and as the mortality was more than 75 percent. (Koeniger 1884, 419; see also Worcester 1908, 10)

Fernando Almeda notes that when cholera struck Cantilan in the province of Surigao in 1882, the 'natives died like animals' and all the priests could do was hear the confessions of a terrified people preparing to die. The local church, according to Fr Martin Juan, became a 'nightmarish sanctuary of men [*sic*] in terror and despair; many of those who went to confess would die before they would reach their homes' (Almeda 1933, 125). Several commissions for public health were established and last-minute quarantine and hygiene regulations were imposed, but to no avail. Large numbers of people were buried in mass graves, and there were instances where gravediggers, while burying the cholera victims, themselves succumbed on the spot (ibid.).

There was a close connection between natural disasters and disease. One observer of the October 1882 typhoon commented as follows on the inadequacy of institutional learning from past mistakes, and the inherently destructive power of the typhoon:

> Let us be just ... The misfortunes which have been afflicting this country is [*sic*] of such magnitude, obstructing its progress, that it hardly has had any respite to recover from the harsh blows of the misfortune ... In the area of urban construction, the best minds have been perplexed in the face of real-

> ity. For the past three years they have rejected buildings of stone rubble-work, replacing the heavy roofs with roof tile with others of galvanised iron, but in two hours the iron blew away and pieces of board and even the walls collapsed, flooding the houses ... The country's nature has such powerful manifestations, while the riches earned at the expense of long and increasing work disappear in a moment through unfortunate accidents, there can be no public prosperity and no lasting general security. The country cannot progress steadily on the road to progress. ('A description of the hurricane which devastated the capital and various provinces of the Philippines on October 20, 1882,' Diaro De Manila, 20)

During the 1888–1889 cholera epidemic, 41 provinces stretching between the Central Luzon Plain and Mindanao were affected and at least 67,612 died—although the Spanish authorities refused to register cholera as a cause of death into the 1890s (ibid., 6, 11). Moreover, during epidemics, cultivation often ceased. For example it was stated, in the 'Report of the Director of the Weather Bureau, 1903,' that the cultivated lands of the northern part of Zamboanga Peninsula were completely abandoned because of a cholera epidemic that was raging in the villages. Hence in May and June, the months normally associated with transplanting rice, surviving farmers were unable to prepare their fields.

In their efforts to eradicate the outbreak, the beleaguered Spanish authorities sought help from the foreign physicians of steamers anchored in Manila Bay, who cooperated with local authorities and doctors to provide urgent help to the thousands stricken with cholera (ibid.). However, the doctors failed to prevent the late nineteenth-century cholera epidemics, and nor did the Spanish engineers build a sufficient number of regional hospitals to adequately cope with the spate of cholera epidemics in the final decades of the nineteenth century (Owen 1982, 205).

American Rule and the Cholera Epidemic of 1902–1904

Following the Spanish—American War of 1898, Spain ceded the Philippines to the United States, provoking a nationalist Filipino uprising against their new colonial ruler from 1899 to 1902. Over 4200 Americans and 220,000 Filipino died as a result of violence, disease and famine. Cholera outbreaks played a central role, causing thousands of deaths from disease and hunger, notably in the stricken war-torn provinces of Leyte and Samar (see 'Numbers of epidemics reported by 54 barrios in Samar and Leyte, 1863–1930' and 'Reports of Epidemics and Hunger/Famine, 1802–1951').

The year 1902 was characterised by cholera and smallpox epidemics and severe typhoons. It also marked an El Niño period which lasted until the end of 1903. In May 1902, cholera arrived in Tacloban, Leyte, aboard a small native trading boat. From Tacloban, it rapidly spread across the island through trade and social intercourse and by boat from Leyte to Basay, Samar and the neighbouring island of Cebu, where in 1902 about 20,000 people died of the disease (Hack 1904, 231). Starving Visayans fleeing the disease were driven to eat the flesh of rotting animals killed in floods and conflicts, despite the risk of contracting anthrax. Again, from March 1902 to 1904, Manila was hit by a cholera epidemic that killed almost 10,000 people, while over 105,000 deaths were recorded in the surrounding provinces (Worcester 1908, 16).

Indeed, for the first ten years of their occupation of the Phillipines, American officials, doctors and scientists blamed deteriorating health and sanitation conditions on the inter-connected impact of the war, weather and microbial pathogens. The health and hygience measures introduced by the United States army were draconian: houses that had been occupied by the sick were burned, and contaminated populations locked down and quarantined. A significant number of American officers and administrators who failed to take adequate precautions against infection also died of the disease.

American scientists and public health officials realised that improvements to sewage systems and sanitation services, food processing standards and personal hygiene would be required to limit future risks of cholera and other water- and food-borne diseases. They also recognised that adverse weather conditions and climate variability would continue to have severe impacts on vulnerable segments of the population, particularly those residing in certain parts of Manila and in towns dotting the exposed east coast of the archipelago, fronting onto the Pacific Ocean, where all Philippine typhoons originate. From early on, American attention focused on Manila's boat-dwelling population (the *Fourth Annual Report of the Philippine Commission 1903* (1904). When cholera occurred on board these craft among a mobile population living in such close proximity to the heart of the city, it quickly spread throughout the lower reaches of the Pasig River. Rather than risk quarantine, Filippinos often threw the corpses of cholera victims into the river at night (ibid., 91–92).

The American administration erected public latrines ashore and public standpipes on the waterfront to give the boat-dwelling population access to clean water (ibid.). Under the remarkable leadership of Dr Victor Heiser (1873–1972) and his Bureau of Science from 1902 to 1922, methods of effective disinfection were developed and advances made in public health and sanitation services and the treatment of cholera and amoebic dysentery (Anderson 2007, 69–73; Gleeck 1976, 147–150). The general decrease in cholera from 1908 to 1912—when not a single case of cholera was reported—was due in large measure to a general improvement in the water supply that followed the development of potable water systems with adequate reservoirs for cities like Manila and Cebu, and increased use of distilled and artesian waters for drinking purposes (*Report of the Philippine Commission to the Secretary of War 1912* 1913, 17, 122). However, other factors, including the introduction of a new sewage system, more effective disposal of garbage and refuse, the destruction of unsanitary slums and watercraft, and the relocation of populations, played their part in controlling the spread of these diseases (ibid., 17). Nevertheless, the antiquated water and sanitation infrastructure continued to pose a health risk, and cholera outbreaks continued to occur. Not until the mid-1920s did work begin on a new reservoir and aqueduct (ibid., 14–15).

Post-Colonial Response to Water-Borne and Infectious Diseases

In the post-WWII era, the Philippines experienced high population growth and remained a nursery for disease, and governments and provincial authorities consistently failed to invest in adequate housing, clean water supplies and proper sanitation services.

Most people, particularly in rural areas, continued to obtain water from force pumps, open wells, rivers or lakes (Vreeland 1976, 136). In Manila and neighbouring provinces, the water and waste-disposal systems regularly broke down due to rising population pressure and the adverse impact of heavy storms and floods. During typhoons and periods of torrential rain, kerbside drains in Manila flooded and streets became rushing streams of filthy water. In low-lying locales and poor districts like Tondo, storm-related flooding could take months to subside. In such situations, cholera, typhoid and other infectious diseases posed a constant threat.

In the 1970s, a devastating series of typhoons and floods led the Red Cross to extend emergency medical and relief aid to families in the hardest-hit provinces. For example, Typhoon Aring, which struck Panay in early December 1976, caused 50 million pesos of damage (approximately USD$1 million) (*Bulletin Today*, 12 December 1976). More than 3000 families were affected by the ensuing floods in Capiz where the Philippine National Red Cross distributed disaster relief assistance such as water purifiers, blankets and medicines to flood victims (*Bulletin Today*, 5 December 1976).

Water-borne and infectious disease forced the Marcos government to develop practical remedial measures to contain the spread, as well as to alleviate the deplorable living conditions of people displaced by typhoons, floods and epidemics. For example, it authorised the use of military helicopters, amphibious vehicles and motorboats to carry emergency stocks of vaccines, drugs and other supplies to storm and disease victims, as well as to deploy full outbreak response teams, including epidemiologists and water and sanitation experts. However, besides using the military and local disaster-coordinating councils, post-Marcos governments have also been forced to call upon international agencies such as the World Health Organization to provide disaster relief and measures to prevent the spread of infectious diseases. Moreover, they have often hidden from the Fillipino population the true reasons for such events. For instance, in November 2001, Governor Pedro Romualdo argued that it was rampaging nature, not illegal logging, that had caused the flash floods that triggered such widespread devastation and loss of human life on Camiguin Island. Again, Typhoon Nanang (6–12 November 2001) caused high human and animal mortality, but the Department of Health denied that there had been any outbreak of disease.[5]

[5] Typhoon Nanang (or Lingling) killed 184 people during the course of its deadly trek across the islands. On Camiguin, a resort island in the southern Visayas, entire villages were wiped out by flash floods and mudslides. In Mahinog town 104 residents died. Nanang's damage came from heavy storm rainfall that triggered flash flooding. The scale of the devastation required preparation of mass graves for those bodies that could be retrieved quickly to aviod the outbreak of a pandemic. In Mahinog the bodies piled up so rapidly that the town ran out of coffins and embalming fluid. Consequently, the dead had to be immediately buried in mass graves because of the possibility of the spread of diseases. See Gomez (2001), Tran (2001), Teves (2001), Alipao (2001).

Conclusion

While the early colonial Philippines did not experience the demographic collapse of Spanish America caused by the widespread introduction of Old World diseases and the trauma of conquest, the impact of tightly-spaced sequential super typhoons was neverthless devasting, notably in terms of the outbreak of deadly diseases such as cholera. Over the next several decades, the impacts of ongoing typhoons by 2020–2030 could, with steady population growth, be catastrophic. Filipinos suffering deprivation in some of the poorest areas of Luzon and the Visayas are becoming ever more vulnerable to such events. Arable land, water supplies, rainforests and food sources are diminishing at an alarming rate while typhoons, floods, drought and infectious diseases are becoming ever more prevalent. This speaks to the urgent necessity of preventative measures in which not only the Fillipino authorites but also the global community will need to participate.

References

"Juan" Death Toll Rises to 9; Diarrhea Kills 10 in Metro Manila. 2002. *Philippine Daily Inquirer*, July 23.

A Description of the Hurricane Which Devastated the Capital and Various Provinces of the Philippines on October 20, 1882. 1882. In *Diario de Manila*, 20. Manila: Ramirez y Giraudier.

Alipao, Tim. 2001. Rescuers Call for Help to Dig Out Dead, Look for Missing from Philippine Tropical Storm. *Associated Press Newswires*, 11/7/2001.

Almeda, Fernardo A., Jr. 1933. *Story of a Province Surigao Across the Years*. Manila: Philippine National Historical Society.

Anderson, Warwick. 2007. *Colonial Pathologies: American Tropical Medicine, Race, and Hygiene in the Philippines*. Quezon City: Ateneo de Manila University Press.

Annual Report of the Governor General of the Philippine Islands 1926. 1928. Washington: Government Printing Office.

Archives of the Manila Observatory (AMO). n.d. Selga Papers. *Estudias sobre las inundaciones de Pangasinan causas y efectos*, p. 579.

Atienza, Paul. 2001. Playing in Floodwaters Banned in Makati. *Today*, August 25, 2.

de Bevoise, Ken. 1995. *Agents of the Apocalypse: Epidemic Disease in Colonial Philippines*. Princeton: Princeton University Press.

Braudel, Fernand. 1981. *The Structures of Everyday Life: Civilization and Capitalism 15th–18th Century*. Vol. 1. New York: Harper and Row.

Bulletin Today, 5, 12 December 1976.

Census of the Philippine Islands: Taken under the Direction of the Philippine Legislature in the Year 1903, vol. III. 1905. Washington, DC: United States Bureau of Census.

Crosby, Alfred. 1972. *The Columbian Exchange, Biological and Cultural Consequences of 1492*. Westport, CT: Greenwood Press.

———. 1986. *Ecological Imperialism and the Biological Expansion of Europe 900–1800*. Cambridge: Cambridge University Press.

Dery, Luis. 1997. *The Kris in Philippine History: A Study of the Impact of Moro Anti-Colonial Resistance*. Quezon City: L.C. Dery.

———. 2006. *Pestilence in the Philippines: A Social History of the Filipino People, 1571–1800*. Quezon City: New Day Publishers.

Doeppers, Daniel F., and Peter Xenos. 1998. A Democratic Frame for Philippine History. In *Population and History: The Demographics of the Modern Philippines*, ed. Daniel F. Doeppers and Peter Xenos, 3–16. Madison: University of Wisconsin Press.

Fourth Annual Report of the Philippine Commission 1903, Part 2. 1904. Washington: Government Office.

Gladstone, Rick. 2013. Ravaged by Typhoon, Philippines Faces Threat of Serious Diseases. *The New York Times*, November 14. http://www.nytimes.com/2013/11/15/world/asia/tacloban-philippines-typhoon-haiyan.html.

Gleeck, Lewis E., Jr. 1976. *American Institutions in the Philippines (1818–1941)*. Manila: Historical Conservation Society.

Gomez, Jim. 2001. President Arroyo Visits Scene of Devastation and Death Left by Storm. *Associated Press Newswires*, 11/11/2001.

Griffiths, Thomas, and Libby Robin. 1997. *Ecology and Empire Environmental Histories of Settler Societies*. Melbourne: Melbourne University Press.

Hack, Charles W. 1904. Report of Special Sanitary Inspection of the Islands of Cebu, Bohol, Samar, and Leyte, by Dr. Charles W. Hack, Special Inspector. In *Fourth Annual Report of the Philippine Commission 1903*, Part 2. Washington: Government Office.

Koeniger, M. 1884. Ueuber epidemisches Auftreten von Beriberi in Manila 1882/83. *Deutsches Archiv für Klinische Medizin* 34: 419–432.

McNeill, William K. 1979. *Plagues and Peoples*. Harmmondsworth: Penguin Books.

Mercado, Juan L. 2001. Typhoons and "Coping with Tyranny of the Urgent". *Today*, November 12, p. 9.

Newson, Linda A. 1998. Old World Diseases in the Early Colonial Philippines and Spanish America. In *Population and History: The Demographics of the Modern Philippines*, ed. Daniel F. Doeppers and Peter Xenos, 17–36. Madison: University of Wisconsin Press.

———. 2009. *Conquest and Pestilence in the Early Spanish Philippines*. Honolulu: University of Hawai'i Press.

Numbers of Epidemics Reported by 54 Barrios in Samar and Leyte, 1863–1930. Historical Data Papers, Visayas. PNL.

Owen, Norman G. 1982. Abaca in Kabikolan: Prosperity Without Progress. In *Philippine Social History Global Trade and Local Transformations*, ed. Alfred W. McCoy and Ed.C. de Jesus, 190–212. Quezon City: Ateneo de Manila University Press.

Philippine National Archive (PNA). 1831. El Gobernador Capitan General de Filipinas a el Junta Superior de Real Hacienda, November 12, Guerra 1831–1833, fol. 171B, 'A Proclamation for Voluntary Contributions to Aid the October 1831 "Baguio" Victims,' exp. no. 489.

———. 1893. Varios Jefes de Provincia a Secretario del Gobierno General de Filipinas, May 7–13, Calamides Publicas (Baguios y huracanes), B2, 1845–1898, expediente no. 124, fols 606–618B.

———. n.d. Alcalde Mayor de Camarines a Direccion General de Administracion Civil, Calamides Publicas (Baguios y huracanes), B4, 1863–1897, exp. 147–48, fols 338–341B.

Quinn, W.H., and V.T. Neal. 1992. The Historical Record of El Niño Events. In *Climate Since A.D. 1500*, ed. Raymond S. Bradley and Philip B. Jones, 623–648. London: Routledge.

Report of the Director of the Weather Bureau. 1903. *Meteorological Bulletin for 1903.* Manila Weather Bureau, 182.

Report of the Governor General Philippine Islands, 1926. 1928. *Annual Report of the Governor General of the Philippine Islands 1926.* Washington: Government Printing Office.

Report of the Philippine Commission to the Secretary of War 1912. 1913. Washington: Government Printing Office.

Report of the Secretary of Public Instruction. 1925. *Annual Report of the Governor General of the Philippine Islands 1923.* Washington: Government Printing Office.

———. 1930. *Annual Report of the Governor General of the Philippine Islands 1928.* Washington: Government Printing Office.

Report of the Secretary of the Interior. 1904. *Fourth Annual Report of the Philippine Commission 1903*, Part 2. Washington: Government Printing Office, 115–117.

Reports of Epidemics and Hunger/Famine, 1802–1951. Historical Data Papers – Visayas. Philippine National Library (PNL).

Selga, Miguel. 1920. El Baguio de 22 Octubre de 1831 en Manila y provincias vecinas. *Revista de la Societad Astronomica de Espana y America*, Vol. xi, No. 78 (17 April).

———. 1935. *Typhoon of 22 October 1831. In Charts of Remarkable Typhoons in the Philippines, 1902–1934.* Catalogue of Typhoons 1348–1934.

———. 1939. Meteorological Note 45. 'Los baguios de Noviembre de 1821 en Samar.' *Meteorological Bulletin*, December, p. 301.

———. n.d. *Accidentes Atmosphericas Consiguados en los Libros Canonicos de la Iglesia y Legajos Historicas del Pueblo Tanay, Rizal*. AMO, Box 8, Item 21, Selga Papers, Meterological Note.

Teves, Oliver. 2001. Lingling Becomes Typhoon, Heads Towards Vietnam as Philippines Counts Its Dead. *Associated Press Newswires*, 11/10/2001.

The Dominicans and the epidemic of cholera at Manila in 1820. AMO, Selga Papers, Box 5, Item 17.

Tran, Tini. 2001. Typhoon Lingling. *Associated Press Newswires*, 11/11/2001.

Vreeland, Neena. 1976. *Area Handbook for the Philippines*. Washington, DC: United States Government Printing Office.

Wachtel, David. 1977. *The Vision of the Vanquished: The Spanish Conquest of Peru Through Indian Eyes*. Sussex: Harvester Press.

Warren, James Francis. 2015. Haiyan's Lingering Aftermath. *Asian Currents, Asian Studies Association of Australia*, February 15. http://asaa.asn.au/haiyans-lingering-aftermath/.

Worcester, Dean C. 1908. *A History of Asiatic Cholera in the Philippines*. Manila: Bureau of Printing.

CHAPTER 6

Malaria in Precolonial Malagasy History

Gwyn Campbell

Introduction: The Advent of Malaria in the Highlands of Madagascar

Although malaria has always been one of the greatest killers of humans in Madagascar, its history has been largely ignored, even by scholars focussing on medical history in the island (Anderson 2017). European accounts had long testified to the presence of malaria on the Malagasy lowlands, but little attempt has been made to trace the origins of malaria in the central highlands. In 1987, Yvan-Georges Paillard asserted that malaria first erupted on the plateau from 1895 due to the social upheaval that followed the implantation of colonial rule (Paillard 1987, 38, 40).[1] A number of subsequent works have challenged Paillard's claim. In one of the most

Research for this chapter received support from the Social Sciences and Humanities Research Council of Canada.

[1] Such claims have a long heritage dating to the first years of the French colonial takeover—see, for example, Blanchard (1907).

G. Campbell (✉)
McGill University, Montreal, QC, Canada
e-mail: gwyn.campbell@mcgill.ca

G. Campbell, E.-M. Knoll (eds.), *Disease Dispersion and Impact in the Indian Ocean World*, Palgrave Series in Indian Ocean World Studies, https://doi.org/10.1007/978-3-030-36264-5_6

recent of these, Eric Jennings (2006, 128–129) underscores the very high incidence of malaria among French invasion troops in 1895, but concurs with Françoise Raison-Jourde (1991, 684–685) that the first epidemic to hit the highlands was in 1878. J. Mouchet et al. (1997) consider this event to have been linked to an inflow of Sakalava from Western Madagascar, where malaria was endemic, to work on church construction, and to the profusion in the highlands of irrigated rice fields which offered a breeding ground for malaria-carrying mosquitoes, notably the *Anopheles funestus* (Mouchet et al. 1997, 123–124). By contrast, Bernard-Alex Gaüzère and Pierre Aubry (2013, 149) associate the outbreak with an allegedly massive influx of Africans to work the irrigated rice fields.

However, a close examination of archival material, notably that emanating from agents of the London Missionary Society (LMS), indicates a more complex story. One of the earliest documented uses of the term "malaria" in Madagascar was by the Scottish surgeon Robert Lyall (1790–1831) in October 1827 (Chapus and Mondain 1954, 54). However, only in the twentieth century, after the process of infection and means of combatting the disease became known, did the word "malaria" become commonly used. Most references to the disease in works dealing with Madagascar prior to the French takeover in 1895 are to "fever", often called by European writers "Madagascar fever". Indeed, so notorious was Madagascar fever, and so well known its symptoms, that, unless otherwise specified, one can be reasonably certain that references to "fever" in Madagascar meant malaria.

European records and Malagasy oral traditions attest to the long-standing existence of malaria in Madagascar where its diffusion was largely determined by climate, geography and human activity. The island's climate varies according to latitude and the influence of the dominant eastern trade winds. In Eastern Madagascar, characterised by a rectilinear coastline with few natural harbours and a coastal plain averaging 50 km in width, tropical conditions prevail. It is wet and humid throughout the year, with an average annual rainfall of 2950 millimetres (mm)—almost double that at Antananarivo on the central plateau—and temperatures varying from an average of 24.3° in the north (Vohimara) to 22.8° in the south (Tôlanaro) (Oliver 1886, 285, 451–452; Sibree 1915, 53; Robequain 1958, 50–51). It possesses a series of coastal lagoons and extensive marshland to the immediate hinterland and, on the interior escarpment, supports a tropical rainforest (Deschamps 1959, 13, 15).

Western Madagascar, which comprises an indented coastline broken by major rivers, and a wide hinterland plain, is affected by three climatic zones. The northwest, strongly influenced by the northern monsoon from October to March, and occasionally visited by a cyclone from the east, has an average annual rainfall and temperature slightly below that of the east coast, although it experiences a hot dry season from April to September. The mid-west, less affected by the northern monsoon and by eastern cyclones, experiences two distinct seasons: hot and dry from April to October and hot and wet from November to March. The temperature there is higher by an annual average of 3° in the north and 2° in the south than at similar altitudes in the east of the island. Average annual rainfall varies from near that of the east coast, in the north, to about 492.5 mm close to Toliara (Tulear), in the south. To the south and interior of Toliara lies a semi-desert that frequently experiences no rainfall (Campbell 2005a, 20–21).

Madagascar's interior comprises an indented plateau, from 800 to 2800 metres high, which runs on a north-south axis almost the entire 1592 km length of the island. A forested escarpment divides it from the eastern lowlands, while to the west the plateau descends more gradually to a wide lowland plain. The high central plateau, 1300 to 1700 metres above sea level, experiences two distinct seasons: hot and humid from November to March and dry from April to November. July and August are particularly cool and windy. Southeast trade winds prevail, although north- and south-westerlies are not unknown. Plateau temperature and rainfall are well below those of the east coast. The plateau is hit often by hailstorms and occasionally, between November and March, by cyclones travelling inland from the east coast (Hugon 1808, 11; Le Sage 1816, 102; Oliver 1886 vol. I, 450 and vol. II, 3; Grandidier 1928, 6–8, 30–39).

Current evidence indicates the probability that the first permanent human settlements in Madagascar, and thus the first possibility for malaria to be present in the island, date to the eighth or ninth century. The first settlers comprised people of both Austronesian and Bantu-speaking origin who migrated to Madagascar from East Africa. It is thus probable that malaria was introduced to the island by these first settlers from East Africa, rather than from Austronesia where, at the time, malaria was absent. This is confirmed by DNA evidence indicating that the dominant sickle cell haplotype in Madagascar (91.4 per cent), which conferred protection against malaria, is of "Bantu" origin (Hewitt et al. 1996). The first human settlers spread along the coasts and lower reaches of rivers that, except for the arid sparsely populated south, proved ideal terrain for the mosquito and its breeding habits. As malaria became endemic in these low-lying belts, it placed the non-sickler Austronesian element at a disadvantage, favouring the survival of sicklers of Bantu-speaking origin and those of mixed Bantu-speaking and Austronesian origin who inherited the sickle cell trait from the African parent.

There were initially few opportunities for malaria to become established in the central plateau of Madagascar because of its altitude, climate and a sparse scattered population. The central highlands lie between 1300 and 1700 metres above sea level, above the altitude (1000 m) conventionally believed to have been the upper limit affected by malaria (Blanchard 1907, 186), and experience a dry season from April to November, that was particularly cool and windy in July and August (Campbell 2005a, 20). However, whereas lowland Madagascar was first permanently settled from the eighth century, human activity becomes noticeable in the plateau interior only from the thirteenth and fourteenth centuries with cultivation of hillsides and swamplands. Around 1500, intensive wet rice cultivation was established in the area around Antananarivo, and thereafter significant population growth occurred. However, the first major drainage and irrigation projects to promote riziculture are considered to have commenced in the last third of the sixteenth century, with more sophisticated irrigated rice-growing techniques being introduced during the eighteenth century in response to three famines: the *Tsimiofy* (lit. "do not peel"), a major seven-year famine in Imerina that occurred during the reign of Andriamasinavalona (r. 1675–1710), possibly from *ca*. 1696 to *ca*. 1703; the *Mavovava* (yellow mouth) in *ca*. 1747; and another, the "younger famine" (which caused less deaths than the *Mavovava*), which, if regional comparisons are taken into account, may have occurred in the years 1755–1756 (Campbell 2019, 54–56).

Direct climatic indicators for the central plateau of Madagascar are sparse, although it probably experienced a more arid dry period from 1300, notably from the 1790s to 1810s. It was also affected by high sulphur-rich volcanic activity from 1693 to 1696 that caused surface temperatures globally to fall by 0.2–0.3 °C, possibly from 1696 to 1699. The impact of volcanism was accentuated by the ENSO effect in 1692–1695 and 1702. Certainly the generalised extreme cold event centred around 1700 is confirmed for South-western Madagascar by coral records and may have been reflected in the *Tsimiofy*. Another cold period started around 1709, and there was a marked ENSO event from 1715 to 1716. Of notable comparative significance are the sulphuric volcanic activities from 1740 to 1744, and 1752 to 1756, and the very strong ENSO effects in 1754 and 1759 (Campbell 2018b).

Such climatic and other environmental events, occurring during the Little Ice Age, probably ensured that highland Madagascar was then too temperate for the mosquito to survive and breed. The main body of settlers in the interior, according to oral sources, comprised Austronesians shipwrecked off the coast of the island who migrated to the interior to escape the "fever" that prevailed on the lowlands. Later known as the Merina, they, unlike lowland Malagasy, had little protection against malaria because they did not possess the sickle cell trait, and only a few who travelled to the lowlands and survived malarial attacks developed resistance through exposure (Campbell 2005b, 6).

This set the context for the early European perspectives on malaria in Madagascar. This is, for example, reflected in the work of LMS director, William Ellis (1794–1872), who remarked in 1838:

> The Malagasy fever, or rather fever and ague together, is called *tazo*. This is the most prevalent and destructive malady in the whole island, especially to the Hovas and Europeans. Ankova [i.e. Imerina], Fort Dauphin, and some of the northern provinces, are the only parts of Madagascar which are throughout the whole year exempt from its formidable ravages. Other parts are exempt at certain seasons; and in some provinces it is so destructive, that certain districts are said to resemble, during the months of December, January, and February, the fabled valley of the deadly Upas, where the whole atmosphere was loaded with poison. (Ellis 1838, vol. 1, 214)

Thus most European attempts to establish settlements in the Malagasy lowlands were doomed to failure. For example, when in 1643 Jacques Pronis (d. 1655) tried to establish a garrison to the hinterland of present-day Fort Dauphin (Tôlanaro), in Southeast Madagascar, fully one-third of his company fell victim to malaria and died (Ellis 1838, vol. 2, 8). Similarly, the French settlers who tried to establish a colony at Foulepointe, in Northeast Madagascar, in 1807, were decimated by malaria (Ellis 1838, vol. 1, 29–30).

By contrast, the central highlands, specifically Imerina, had a reputation for being malaria-free. As Ellis noted:

> The great elevation of the province of Ankova [Imerina], perhaps five or six thousand feet above the level of the sea, the absence of forests, the general dryness of the soil, the partial extent to which luxuriant vegetation is spontaneous, and the cultivation of many of the marshy parts of the soil, will be sufficient to account for its salubrity. The weather on the coast is generally hot and damp, or rainy; but in the interior the rains are periodical, and, in a great measure, regulate the divisions or seasons of the year. (Ellis 1838, vol. 1, 29–30)

However, many Europeans who travelled to Imerina in the early nineteenth century perished of malaria. For example, when Bibye Le Sage (d. 1843) led an expedition from Mauritius to Antananarivo in 1816–1817 only five of his 32-strong party survived longer than six months, and of those only one remained healthy (Lewis 1835, 237). As Ellis stated (1838, vol. 2, 157–158):

> The fact is, the proper season and mode of travelling up the country were not at that time understood. It is now found that persons, who have lived some time on the coast, enjoy better health there during the rainy season, than if they retire into the interior. Persons newly arriving in the island, consequently not acclimated, if they venture to travel during the rainy season, are then liable to the dreadful disease, which generally proves fatal to strangers remaining under such circumstances at the capital, although they may have enjoyed after their arrival a few days of health. Experience has shown that a period of about ten days forms a crisis. Those who reach the capital, and pass ten days without an attack, may consider themselves safe, as the fever has invariably been found to manifest its symptoms within that period, if at all.

As the nineteenth century progressed, notably with a significant influx of European missionaries and traders from 1862, knowledge of the topography and climatic zones of Madagascar, as of the geography, transmission and treatment of malaria, steadily improved. Thus by the end of the century, it was known that while malaria was endemic throughout most of the lowlands, where the local indigenous population was largely immune to it, it was also endemic in certain less elevated highland zones such as Vonizongo and Antsihanaka. As missionary doctor C.F.A. Moss noted in 1895:

> The north-eastern shores of the island near Vohimarina are comparatively healthy, but fevers are prevalent at the new French port of Diego Suarez in the north. The more elevated parts of the south coast suffer but little from fever. Fever is endemic along the east and west coasts and in the islands near to the shore. Both on the coasts and on the islands it manifests a high degree of intensity, especially among Europeans and the Hova [Merina] belonging to the central province of Imerina, when they visit the low country. The coast tribes, however, enjoy a comparative immunity from the disease.
>
> The French island of Ste. Marie, on the east coast, deserves the evil repute which it has obtained since its first occupation.
>
> The forest zone is not exempt from malaria, which is met especially in the humid valleys, such as Beforona, which are more or less shut in by mountains. The valley of Angavo, again, although at an altitude of 3000 feet [914 m] is excessively malarious; the natives, who are here mostly of Hova origin, suffer severely from the malarial cachexia.
>
> In the bare, open, central province of Imerina, at an elevation of 4000 to 5000 feet [1219 to 1524 m], as well as in the Betsileo country to the south, fever is not endemic; but to the west of Imerina, in the Vonizongo district, where the elevation is less, and the country level, grassy, and in parts marshy, almost every one suffers from enlargement of the spleen. This organ is not unfrequently found to stretch across the abdomen to the right iliac crest. The malarial cachexia is very general here, while frank attacks of fever are rare. The Antsihanaka country to the north of the capital, especially in the neighbourhood of the Alaotra Lake, is highly malarious. (Moss 1895, 331)

The missionaries also knew well the reputation of Angavo, a mountainous region in the eastern forest. In the first phase of the LMS mission to Madagascar, Thomas Rowlands (c. 1804–1828) had in 1827 moved to Ifody, near Angavo, some 80 km east of Antananarivo, with the aim of promoting there the large-scale cultivation of cotton and of raffia and hemp (to produce sail cloth for which he anticipated a larger market). However, he caught malaria in March 1828, while returning to Antananarivo through extensively flooded ground, and died shortly afterwards (Campbell 2012, 708; Ellis 1838, vol. 2, 392).

Again, of Vonizongo, a large district centred northwest of Imerina, one of the first LMS missionaries planning to settle in the region commented in early 1871: "we began to make preparations for proceeding to Vonizongo; but to this the doctors very strongly objected. They said I was quite unfit to face the fever of the west, adding that I little knew to what I was going, and in this they were perfectly right" (Matthews 1904, 88).

That year he did visit the southern portion of the region, but noted: "We were not able to visit the Ankazobe district, at the north end of the province, as malarial fever was raging there that season" (ibid., 95). Indeed, he called Vonizongo "one of the worst fever districts of the island" (ibid., 208). The disease was also endemic in Vakinankaratra, southeast of Imerina, and around Lake Alaotra, in Antsihanaka (ibid., 208). There it attacked Sihanaka and Merina alike, notably from December to March, although some European doctors later that century found it difficult to tell whether enlarged spleens (*atodi-tazo*) were due purely to malaria or to syphilis (Mackay 1893, 52). The LMS missionary Joseph Pearse, who lived in Antsiahanaka in the 1870s, characterised the region as "fever-saturated" (Moss 1913, xiv).

Indeed, it was traditional policy for the Merina court to send high-ranking criminals to such areas of endemic malaria where they almost invariably perished of the disease (Hastie 1822; Ellis 1838, vol. 1, 214–215). In 1857, for example, 57 Christians were condemned to life in irons and banished to the notoriously malarial areas of Ambohiboahazo, Ambatondrazaka and Ankazonamoizana. Of these, only 22 returned alive to Antananarivo following the death of Ranavalona I (r. 1828–1861) (Razafimahatratra 2005, 95–96).

Indeed, an analysis of the archival evidence indicates not only that malaria had long been endemic in highland regions such as Vonizongo and Antsihanaka, as in Angavo, all less than 100 km from Antananarivo, the capital of Imerina, but also that it erupted periodically in epidemic form in Imerina from the 1820s and Betsileo from the 1870s. The following sections of this chapter outline the changing geography of malaria in the central highlands of Madagascar in the imperial Merina era, 1795–1895, and in so doing, explore the timing of malarial outbreaks in the highlands, their cause and their impact.

Causes of Malaria

Today, it is known that malaria is caused by *Plasmodium*, a type of protozoan, a single-celled organism which experiences an asexual cycle, that occurs in the liver and red blood cells of vertebrates, including humans, which, when ingested by blood-sucking mosquitoes, transforms into gamete, a sexual form which completes the life cycle of *Plasmodium*. Gametes produce sporozoites, a cell type that, when feeding, the mosquito injects into the vertebrate bloodstream. The sporozoites accumulate and multiply in the vertebrate liver to produce merozoites—a form that first invades, then in a form of asexual reproduction, replicates and destroys red blood cells (Miller [2018]).

However, the true cause of malaria only became known from the very end of the nineteenth century. The traditional European view is that it was caused by rotting vegetation in stagnant waters. Thus, from a report compiled by LMS missionary William Milne (1785–1822) on Mauritius in 1812–1813, fellow missionary John Campbell (1766–1840), stationed in South Africa, noted: "Behind the town of Tumetave [Tamatave], on the east coast, there is a vast morass, from whence unhealthy vapours proceed, which contaminate the air" ([Milne] 1815). More specifically, William Ellis (1838, vol. 1, 28–29) wrote:

> The malaria which engenders the destructive fever, is supposed to arise from the decomposition of vegetable substances in contact with stagnant water. The mouths of many of the rivers are choked up with sand, so that their waters either pass sluggishly into the sea, or, when not swollen by rains falling in the interior, present the aspect of a broad, unruffled, stagnant lake, for several miles inland. The brackishness of the water, and the absence of crocodiles, often indicate a level below that of the waters of the sea, while much of the ground on the inland side of the bank of sand that is raised along the border of the sea, being below the level of the ocean, extensive morasses occur in several parts of the coast. Many of the lakes are also shallow, and receive large quantities of vegetable matter, furnished in all the rank luxuriance which the heat and humidity of the climate unite to produce; and some of these sheets of water, from the trees and shrubs that grow around, and rise in different parts of their surface, bear a greater resemblance to insulated forests than ordinary lakes.
>
> The effluvia arising from the lakes and swamps near the coast, is extremely prejudicial to health; and by incautious exposure to this, either early in the morning or late in the evening, the fatal seeds of the Malagasy fever may be so deeply received into the human constitution as never to be eradicated.

In 1843, British Rear Admiral John Marshall (1785–1850) of HMS *Isis* ascribed similar causes for malaria when visiting the French-held islands of Sainte-Marie and Nosy Be, off the northeast and northwest coast of Madagascar, respectively. Of Sainte-Marie, he commented:

> The vegetation is luxuriant beyond conception, but the undulating surface of the ground formed of continued ranges of hills all over the island, rising to a Height of from 100 to 400 feet, occasions hollows and flats in which the rain lodges, and acting upon the decayed vegetable and animal matter, materially produces the seeds of the dreadful fever which are here so abundantly sown. The French have done nothing to remove the cause of this dreadful scourge, thus leaving it to triumph over European science, industry and humanity. (Marshall 1843)

European susceptibility to malaria led Radama I (r. 1810–1828) to consider the disease a cornerstone of his defence policy against European aggression. Thus Robert Lyall, Hastie's successor as British Agent to Madagascar, stated after malaria decimated troops participating in the French attack on the northeast coast of Madagascar in 1829: "General Death, as the late Radama well named that mortal disease, the endemical fever of the Island, is fighting the Malagashes' Battles more effectually than cannon, musquets, or sagayes" (Lyall 17 April 1830). It is noteworthy that from 1862 to 1880, fully 99.31 per cent of hospital cases in the French hospital on Nosy Be were malaria related (Moss 1895, 332). When the French seized Diego Suarez, a port in Northern Madagascar, in 1885, malaria was also the major illness encountered amongst their troops (Graph 6.1).

In the late nineteenth century, it was the reluctance of French authorities to administer quinine to their troops that caused such a high mortality rate (estimates vary between 32 and 72 per cent) during the 1895 French military expedition to Madagascar (Jennings 2006, 128–129; Campbell 1992; Cohen 1983; Blanchard 1907, 185, fn. 3).

Symptoms

In Madagascar, the malarial cycle generally commenced from the start of the rains in November, which afforded the female *Anopheles* mosquito ideal sites to lay eggs. In non-sicklers, clinical symptoms of the disease usually started to show in the victim from one to four weeks after the onset of the rains and peaked in the period between March and early May. Symptoms generally included headaches, high temperatures, fevers and chills, but those affected could also experience loss of appetite and energy, dizziness, body pains, diarrhoea and cramps (Hlongwana et al. 2009). Reflecting the experience of early missionaries to the island, Ellis (1838, vol. 1, 215–216) commented:

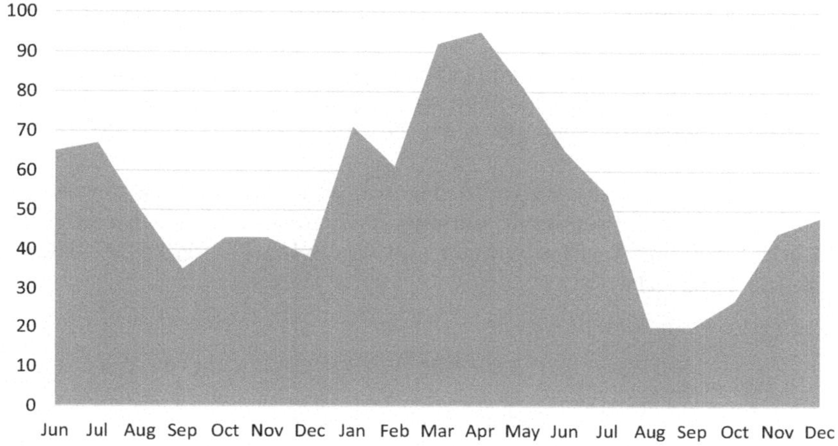

Graph 6.1 Diego Suarez hospital: Malaria as percentage of all cases, 1886–1887. (Source: Moss 1895, 332)

The symptoms of the tazo, or Malagasy fever, vary considerably in different individuals. In some cases its early symptoms resemble those of a violent inflammatory disorder. This is always considered its worst appearance. In others it assumes the form of a remittent, and afterwards an intermittent fever, attended with chills and shiverings. This is regarded as its most favourable appearance, and that which encourages the greatest hopes of recovery. When the former symptoms are strongly marked, there is always great danger; but when the latter only are present, recovery is generally expected.

The symptoms of the fever, in its inflammatory state, are a severe head-ache, languor, pain in the eyes, especially on being turned upwards, dry and hot skin, vomitings, pain in the right side, great thirst, quick and hard pulse, a very foul furred tongue, aversion to food, flushing of the face, restlessness especially at night, loss of sleep, sometimes dread or fear, anxiety, hypochondriasis, delirium, an apparent improvement, and then death suddenly and unexpectedly.

In its most unfavourable appearance, the symptoms are colds and chills in the back, shoulders, and loins, stretching, lassitude, and a disposition to lie down, with great fatigue on the least exertion, a feeling of debility, anxiety, loss of appetite, thirst, uneasiness of mind, rigours, and sometimes stupor. This form is at first remittent, and soon changes into an intermittent fever, which is divided into three stages of cold, hot, and sweating fits in a severe degree. The sensation during the shivering fit is like being pierced with darts. This is followed, if it continues long, with hypochondriasis, dyspepsia, hepatitis, and dysentery; and unless the fever ceases with these symptoms, it speedily afterwards terminates in death.

And as LMS missionary Joseph Pearse commented of Antsihanaka in 1884:

> One of the marked effects upon people born and living in fever districts in Madagascar is that the spleen becomes permanently enlarged, and I have seen cases in which that organ has reached an enormous size. (Pearse 1884, 18)

Almost all foreign visitors to Madagascar were attacked by malaria of which many died. Prominent amongst its victims were Scottish surgeon Robert Lyall and Austrian voyager and writer Ida Pfeiffer (1797–1858) (Pfeiffer 1881, 319).[2]

Treatment

Indigenous Treatment

There were a variety of indigenous treatments of malaria. In Imerina, as a preventative measure, it was common for sufferers to recourse to *sampy*, national talismans believed to possess awesome powers. In epidemics, the *sampy* Ramahavaly was summoned to perform the *miafana* ("to avert") ceremony whereby *ranomafana* ("the water of averting") was sprinkled over an assembled population in order to protect them. Supplicants had to respect Ramahavaly's taboos (*fady*) that included the killing of serpents, serving boiled greens in a house and stirring rice while it was being cooked (Campbell 2012, 543). Less important *sampy* that could also protect against tazo and other illnesses included Rafantaka and Rakapila (Ellis 1838, vol. 1, 413; Campbell 2012, 544).

Once attacked by malaria, sufferers could consult a *mpsikidy*, or diviner. As any disease with nervous symptoms, such as fever, was considered a likely sign of being afflicted by *vazimba* (reputedly the first inhabitants of the Malagasy highlands), the mpsikidy might direct the patient to appease the vazimba through sacrifices and offerings (Radaody-Ralarosy 1971, 66). Other counsel was directed to physical concerns. As Ellis (1838, vol. 1, 217) commented:

> When a person is seized with the fever, the remedy is directed by the sikidy, or divination. Inquiry in such cases is made of the sikidy, in which house the patient must dwell. Then they make his couch, that he may lie on the west of the hearth, near the fire, and administer plenty of rice to eat; yea, they compel him to swallow boiled rice, or any food, as they believe this to be essential to his recovery.

[2] Pfeiffer died of malaria in Vienna in October 1858.

> The most beneficial remedy in the early stages of the disorder is supposed to be rice-water, which produces perspiration, and is supposed to nourish the system during the season of aversion to food. When the skin is dry and hot, or a fit of fever comes on, the vapour-bath is used; leaves, supposed to possess medicinal qualities, being first boiled in the water. This diminishes the force of the fever, and sometimes checks it entirely, if it be used half an hour before the regular time of the appearance of the paroxysms which come on every other day. When the effect of the bath is thus salutary, the patient will then have an interval of ninety-six hours in which to recruit his strength, instead of only forty-eight. He will thereby be proportionably better fortified against the next attack. Besides rice-water, an infusion of herbs possessing aperient qualities are administered; to those they sometimes add a decoction of leaves, which is exceedingly bitter, supposed to act as a tonic.
>
> In addition to the use of the vapour-bath half an hour before the paroxysm comes on, they take the warm bath every evening, which, with a quantity of warm drink, never fails to produce moisture on the skin, and gives the patient rest in the night.

He stated further:

> a vapour-bath is a favourite remedy with the sick, and frequently in the early stages of the fever it is most successfully applied… The Malagasy seat the patient over a large earthen or other pan containing water, spreading over him several large native cloths, and then produce the quantity of steam required by casting pieces of iron, or stones heated red-hot into the water. (Ellis 1838, vol. 1, 222–223)

Most indigenous healers treated malaria with decoctions or infusions from a number of bitter plants, Ellis (1838, vol. 1, 222) commenting that:

> Barks, gums, leaves, roots &c., possessing an aperient, cathartic, diuretic, tonic, or sedative property, are generally applied in cases in which they are specifically required. Hence they are able sometimes to arrest the progress of the fever, when the symptoms of inflammation are violent and decisive. The remedies taken internally consist of decoctions or infusions. External applications are in the form of fomentations, poultices, or ointments made by heating the fat of animals.

It appears that the plants most commonly used by traditional healers to treat malaria were those of the *Zanthoxylum* sp. (Rutaceae family) (Randrianarivelojosia et al. 2003; Parker 1881, 78, 80; Baron 1878, 111, 115). Philippe Rasoanaivo and colleagues note 229 species of plants, 30 per cent of which are endemic to Madagascar, used by indigenous healers to treat the disease. Of these, the most efficacious against virulent contemporary strains of malaria include *bemaifaitra* (*Cassinopsis madagascariensis*), *hazomafaitra* (*Samadera madagascariensis*), *fatray* (*Evodia fatraina*) and *afatray* (*Urophyllum lyallii*) (Rasoanaivo et al. 1992).

Occasionally, Malagasy patients would request quinine from Europeans. However, many preferred to eschew both traditional and Western medicines and adopt the *vita'ny tazo* approach—the belief that if they survived an initial attack of malaria, they gained immunity against it ([Pearse] 1884, 18; Mackay 1893, 52–53; Matthews 1904, 96; Carayon 1845, ix).

Western Treatment

European treatments also varied. When André Coppalle was first attacked by malaria near Foulepointe in August 1825 and became delirious

> A trader, seeing me in this state, immediately obliged me to take an emetic. It took little time to have an effect, during and following which, I felt a burning fever and emotions which permitted me to enjoy only a part of my intellectual faculties. A second emetic, administered at midnight, made no noticeable change in my situation.
>
> On the 10th, a large dose of Le Roy's purgative caused numerous excretions, and, with the aid of a small vesicatory affixed to my arm, it largely dissipated the terrible pains I felt in my head; I still had a fever; but my faculties had fully returned.
>
> I spent the night of the 10th in extreme agitation, which increased during sleep; I felt inside a burning heat that made me move from one part of the apartment to another, in search of some coolness.
>
> On the 11th, two ounces of Glauber's salt taken successively produced no effect.
>
> On the 12th, 13th, 14th and 15th, they gave me Le Roy's purgative. During this space of time, the fever gradually diminished in intensity, but did not leave me entirely until the 16th, and left me so weak that I could hardly support myself, even with a stick. This weakness lasted until the 20th, when I began again to find some flavour in food. Finally I thought I was well enough today to undertake the long journey to Imerina. (Coppalle [c. 1827], 14–15)

James Hastie (1786–1826), British Agent at the Merina court from 1820 to 1826, gained a reputation for the successful treatment of malaria among both the Merina and Europeans using imported salts and barks (Hastie to Barry, Port Louis, 18 Apr 1822a, HB 7, NAM), while Ellis (1838, vol. 2, 216) commented in 1838:

> If the patient be robust, and has a hard quick pulse, Europeans have recourse to bleeding, and other means of reducing the system. When there is much thirst and a hot dry skin, an anodyne antimonial draught is sometimes found useful. If after this no change for the better be observed, mercury is administered, so as to produce ptyalism as speedily as possible, which in general diminishes or removes every unfavourable symptom. Tonics, such as cinchona bark or the sulphat of quinine, are afterwards administered, and the patient recommended to drink plentifully of rice-water, or some acidulate beverage.

Others advocated the use of *Eucalyptus globulus* ([Pearse] 1884, 19), but as the efficacy of quinine became increasingly known, so did its popularity. The French, for example, were using what they termed "kina" (quinine) to combat malaria on their colony of Sainte-Marie in the 1820s, and in 1844 Désiré Laverdant stated: "One protects oneself from harm by sobriety and by the use of quinine sulphate" (Laverdant 1844, 26).

Expansion of Malaria in the Highlands

The major reasons for the expansion of malaria in highland Madagascar were climatic change and the impact of Merina court policies, notably imperial expansion within the island, and *fanompoana*, or unremunerated forced labour for the state. Meteorological conditions directly affect vector reproduction and mortality rates and indirectly influence malarial infection through their impact on blood-feeding frequency and the incubation period of the pathogen (Hay et al. 2000, 9335). Comparatively higher temperatures and little seasonal variation in the tropics are conducive to high mosquito reproduction rates. In common with most of Eastern Africa, Madagascar experienced two decades characterised by lengthy periods of drought, broken by intervals of exceptional rainfall, and diminished harvests, from about 1820 to 1840, and a generally significantly warmer and wetter period from around mid-century to 1880. These climatic events created an environment more conducive not only to the *Anopheles gambiae*, the main malaria vector in the Malagasy lowlands, but also to the *Anopheles funestus*, which became the main malaria vector in the central highlands (Rakotoson et al. 2017; Hay et al. 2000, 9335; Campbell 2018a, 21).

For example, in 1821, when a Merina expedition comprising 70,000 to 80,000 people, of whom only 1000 were trained troops—most of the remainder were slaves and servants—was launched against the Boina Sakalava, between 25,000 and 30,000 died of malaria and famine (Guillain 1845, 58, 60). Radama I thereafter formed a disciplined corps of 13,000 troops accompanied, according to French explorer Charles Guillain (1808–1875), by 7000 porters (Guillain 1845, 64). However, they too suffered such mortality from malaria while on military expeditions in Western Madagascar that Hastie informed the Governor of Mauritius that Radama I's plan to place garrisons on all parts of the coast in 1824 could not be achieved that year (Hastie to Cole, Marouvoie, 14 Aug 1824a, HB 5, NAM)—Hastie emphasising that "The fever is the only enemy to contend with here" (Hastie to [Barry], Marouvoie, 14 Aug 1824b, HB 5, NAM). Indeed, Radama I also contracted malaria on that expedition (Hastie to Lowry Cole, Tananarive, 10 Dec 1824c, HB 5, NAM). It is possible that such experiences made Radama amenable to the advice of Robert Farquhar (1776–1830), Governor of Mauritius (1810–1823), that he drain the swamps to the immediate hinterland of the northeast coast (Hastie to Telfair, Tananarive, 8 Oct 1821, HB 21, NAM): In 1827, Radama ordered an 800-man-strong corvée to transform the littoral lagoons into a coastal canal system (Raombana 1853a, 263—AAM) (Maps 6.1 and 6.2).

In all, an estimated 25 to 50 per cent of Merina soldiers in lowland provinces died each year, mostly of malaria, as did about 160,000 Merina soldiers (close to Raombana's estimate of 150,000 for the years 1820–1853) in imperial campaigns between 1816 and 1853, giving an average of about 4500 soldier deaths a year (possibly 0.8–4.5 per cent of the Merina population). The pace of military campaigning slackened radically from the early 1850s, so that an estimate of 235,000 Merina military dead, mostly from malaria and famine, for the entire period 1800–1895 would appear reasonable (Chart 6.2; Campbell 1991, 432).

Secondly, although the incidence of malaria amongst Merina troops may have declined in the provinces from 1850, it increased on the plateau. The major reason for this was increased state *fanompoana*, which resulted in the mass circulation of forced labour units, military and civilian—a feature often ascribed only to European colonial regimes—between the plateau and lowland and other malarial areas (Campbell 1991, 432–433). Welsh missionary David Griffiths (1792–1863) wrote in 1840 of Imerina:

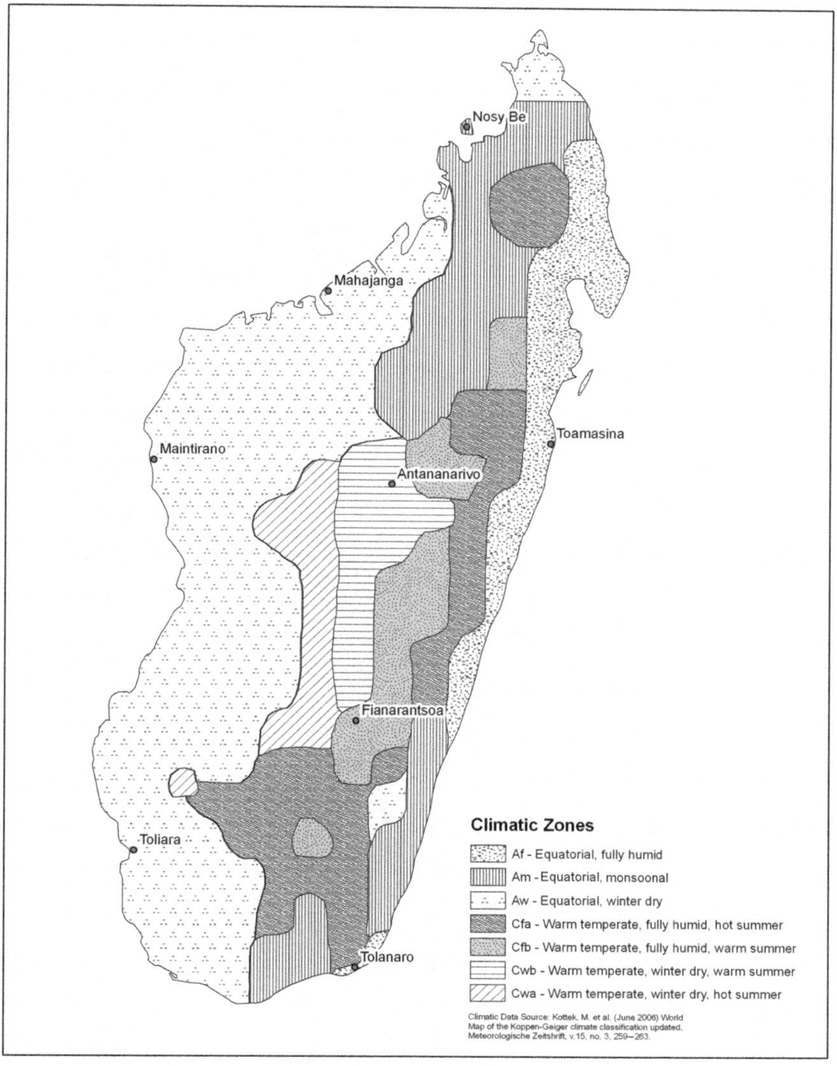

Map 6.1 Madagascar. Contemporary climatic zones. (Source: Drawn by Carl Hughes, IOWC)

Map 6.2 The Merina Empire. (Source: From Campbell 1987, 396)

> The six provinces in the interior of the island are oppressed to the extreme, both soldiers and civilians being compelled to work at a moment's notice,—the most cruel system of slavery ever known! All the people in the inland provinces, and on the eastern coast from Vohimarina to Fort Dauphin, have not a week that they can call their own to cultivate their ground or provide for their families, but are required to engage in some government service or other, as tilling the ground, felling timber, making and carrying charcoal, collecting wax and gum copal, etc., and carrying hides from the interior to Tamatave. All the tailors have their service exacted in the same manner by the Government without remuneration. The people often remark, with feelings of stoical indifference, "We shall not teach our children *any thing*, for the more they know, the harder will be their service!" (Griffiths (1840) quoted in Cousins 1895, 344)

The carriage of raw materials to industrial centres in Imerina was generally allocated to *fanompoana* units of conquered provincial peoples, notably the Betsileo, Bezanozano, Betsimisaraka and Sihanaka (Campbell 2005a, 83). As the homelands of the last three of these ethnic groups were characterised by endemic malaria, they could have transferred malaria to previously malaria-free zones. Additionally, following the 1821 Merina military campaign in Western Madagascar, returning soldiers and their prisoners almost certainly brought malaria back to the plateau interior, which in consequence started to experience severe malarial epidemics (Campbell 1991). Thus in January 1822, Radama I summoned Barnsley, Assistant British Agent to Madagascar, to accompany him to administer medicine to two of the Merina elite who were very ill with fever, the chief of a village near Antananarivo and Radama's sister, wife of Prince Ratefy. There is no evidence that either had travelled to the lowlands—and Radama's sister died a few days after the visit (Barnsley 1821–1822, HB 7, NAM). Barnsley was himself obliged to leave Imerina later that year due to repeated attacks of fever (Hastie to Farquhar, 14 Oct 1822b, HB 7, NAM).

In 1831, the missionaries also reported that malaria was endemic in the region about 100 km west of Antananarivo and that it killed many people (Ellis 1838, vol. 2, 461). It was probably in 1831 that an outbreak of malaria killed the wife and children of Rainitsiandavaka, guardian of the sampy *zanaharitsimandry*, who lived near Mangatany, 50 km to the north of Antananarivo. The tragedy led Rainitsiandavaka into conversation with a Malagasy Christian, Rainitsiheva, and in 1832 to visit the missionaries in Antananarivo and convert to Christianity. In 1834, he marched from his home region to Antananarivo where, at the head of 200 disciples, he proclaimed the end of the world, resurrection of the dead, universal peace and the common ancestry of all people. Freeman and Johns assert that it was the last claim that sealed his fate, as the Merina court would not entertain the idea that they and imported African slaves shared a common ancestry. Rainitsiandavaka and three of his leading disciples were arrested, placed head down in a rice pit, had boiling water poured over them and were then buried. The *tangena* poison ordeal was administered to a further 70 of Rainitsiandavaka's disciples, 18 of whom succumbed. The remainder were sold into slavery and their property confiscated. This event, Raymond Razafimahatratra argues, was the catalyst for the court turning against Christianity (Freeman and Johns 1840, 91–97; Razafimahatratra 2005, 64).

The previous year (1833), David Griffiths (1843, 64) noted: "a large number of the most God-fearing officers and soldiers, many of whom were amongst the most zealous supporters of the cause in the island, perish in military campaigns… they died either of fever or of smallpox". It is noteworthy that 1833–1834 was one of the wettest years in living memory and was followed by a number of fatal cases of malaria in Imerina during the allegedly non-malarial season from May to September (Freeman 13 Oct 1834, HB 9, NAM).

The expansion of malaria into the highlands was accelerated by changes in demographic and settlement patterns. The imposition of security by the Merina state, from the end of the Merina civil wars in *c.* 1795 in Imerina and from *c.* 1820 in Betsileo, induced plateau peoples to leave their fortified hilltop encampments for settlements in the cultivated valley bottoms. This move from a theoretically mosquito-hostile to mosquito-friendly environment was of little importance until the 1820s, from when absorption into, or flight from, *fanompoana* led to the abandonment of large stretches of the irrigated riziculture network. As will be seen from the chart below (Chart 6.1), the greatest burden of skilled and unskilled *fanompoana* was experienced in the 1820s during the reign of Radama I, and especially in the 1830s under his successor Ranavalona I. Following the rice harvest, it was customary in plateau regions to drain the fields of water, but increasingly from the late 1820s, due to labour shortages, water was left to stagnate. Stagnant water in turn attracted the *Anopheles* malaria vector (Campbell 1991, 433).

The abandonment of settlements increased dramatically in the 1830s. Thus by 1840, many formerly cultivated regions in and around Imerina were abandoned. Charles Campbell, a British military officer who had visited the region in 1827, noted in his journal for 17 June 1840, travel-

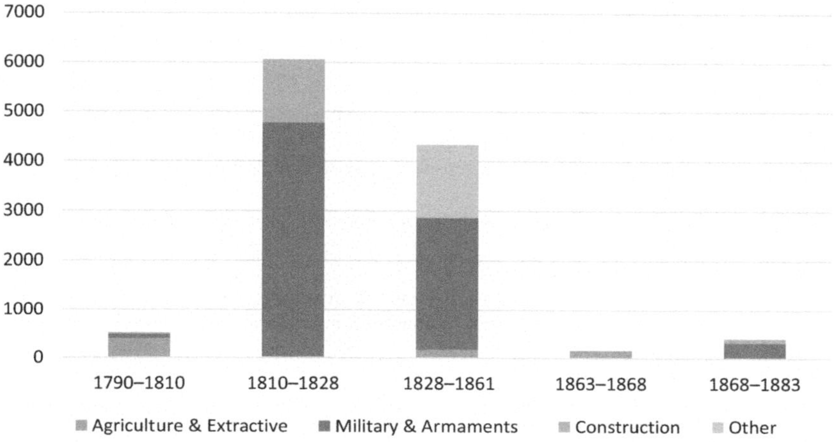

Chart 6.1 Skilled fanompoana. (Source: Campbell 2005a, 126)

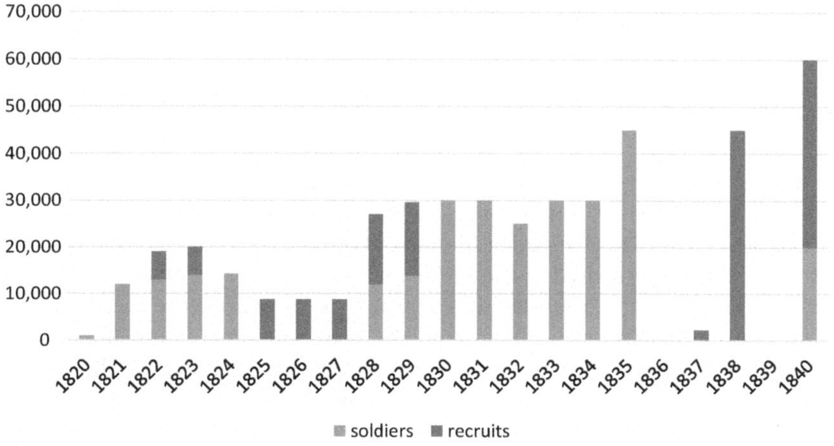

Chart 6.2 Estimated growth of imperial Merina army, 1820–1852. (Source: Campbell 1988a, 470)

ling to Imerina from Tamatave (Toamasina), that "Rice fields formerly in cultivation abandoned and overgrown with weeds and high grass" (Campbell 1840 HB 14, NAM) characterised the Mangoro valley to the east of Antananarivo (Charts 6.1 and 6.2; Table 6.1).

Table 6.1 Highland Madagascar: Malaria, 1817–1896

Year	*Event*
1817–1853	On imperial Merina army expeditions to lowlands, 50% die of malaria
1822	Imerina: fever [?] (starts January)
1826	Imerina: starts in west—extensive disease with high mortality [malaria?] (starts in May)
1833–1834	Imerina: rainfall highest in living memory—rice harvest spoiled (November–February)
1834	Imerina: many malarial cases
1863	Imerina: malaria; smallpox epidemic; choreomania (*ramanenjana*) (Feb–Apr) started "west or south-west of Imerina" and by March, common in Antananarivo
1864	Imerina: malaria
1874	Particularly severe outbreak of malaria in Vonizongo
1877–1880	Highlands: malaria epidemic—kills many; Betsileo: smallpox
1882–1883	Imerina: typhoid and malaria—severe
1884	Smallpox epidemic, measles, severe outbreak of malaria, choreomania
1890	Region around Fianarantsoa: malaria epidemic in May
1891	Highlands: drought, famine, malaria, smallpox
1895	Highlands; malaria epidemic
1896	Imerina and Betsileo: malaria

Sources: IOWC database; Freeman 13 Oct 1834, HB 9, NAM; Matthews 1904, 100; Moss 1913, 142, 164–165

From 1853, military expeditions dropped off dramatically. Consequently, *fanompoana*, a major reason for malarial outbreaks in Imerina in the period 1821–1853, eased, and malarial outbreaks in Imerina decreased. This trend was confirmed in the 1860s, with the adoption of a more liberal commercial policy by the Merina government. However, the creation of a state-church in 1869 radically altered the situation. Thereafter, missionaries became imperial agents and missionary chapels and schools the prime institutions for the summoning of *fanompoana* labour which formed the chief resource of the imperial economy. Missionaries often gave active support to the recruitment and even supervision of *fanompoana*. As *fanompoana* included any service demanded by a superior of an inferior, state-church personnel, from missionaries to school teachers, used their official status to demand a wide variety of *fanompoana* services. *Fanompoana* formed the basis of the imperial economy, which depended even more heavily upon forced labour from the late 1870s due to the financial exigencies of the Franco-Merina War of 1882–1885 and the commercial depression of the 1880s and early 1890s (Campbell 1988b).

From the 1870s, Betsileo, a central province to the south of Imerina that the Merina exploited as conquered colony, felt the brunt of increased *fanompoana*. First came *fanompoana* for the expansion and maintenance of the state-church as an institution. As state-church agents from 1869, missionaries gained full access to *fanompoana* labour which was by definition unremunerated and could be summoned whenever required (Sibree 1870, 530). One of the first components of such *fanompoana* was the building of churches and schools (often held in churches) to cope with the largely compulsory mass conversion to the state-church. Indeed, the building of church property became euphemistically referred to as *'ny fanompoany an 'Andtr'* ("*fanompoana* for God") (Thorne 1888, 43). By 1880, the LMS possessed in Madagascar, predominantly in Imerina and Betsileo, 1024 chapels and 862 schools, while by 1881 the Jesuit mission comprised 228 mission stations, 144 chapels and 170 schools (Campbell 1988b, 64). Following the 1882–1885 Franco-Merina conflict, the only increase in state-church building was in subjugated provinces of the Merina Empire, chiefly Betsileo (LMS 1879, 23; 1882, 41, 45, 90, 102). This was reflected in the fact that, in the 1880s, the number of scholars registered in LMS schools increased in Imerina by 35 per cent and in Betsileo by 200 per cent. The enormous growth in pupil numbers outside Imerina indicates the decision of the Merina court to concentrate the impact of *fanompoana* in Betsileo (Chart 6.3).

The cost of chapel building, in labour and money, placed a heavy burden on peasant communities already afflicted by other *fanompoana*, particularly as most communities were pressurised into constructing and maintaining at least two rival church establishments. State-church *fanompoana* in Betsileo proved so arduous during the building programme of the 1870s and 1880s that many local craftsmen fled, obliging Norwegian missionaries there to import skilled workers from Imerina (Campbell 1988b, 66). Further, as most church buildings were built with a European-style low-pitched roof in contrast to the traditional Malagasy high-pitched roofs, they proved vulnerable to the torrential rainstorms that swept the plateau in the rainy season and were subject to constant repair work (Campbell 1988b, 65). *Fanompoana* labour could not produce all the materials required so that, on average, an ordinary small clay and thatch church cost \$60–125 in pur-

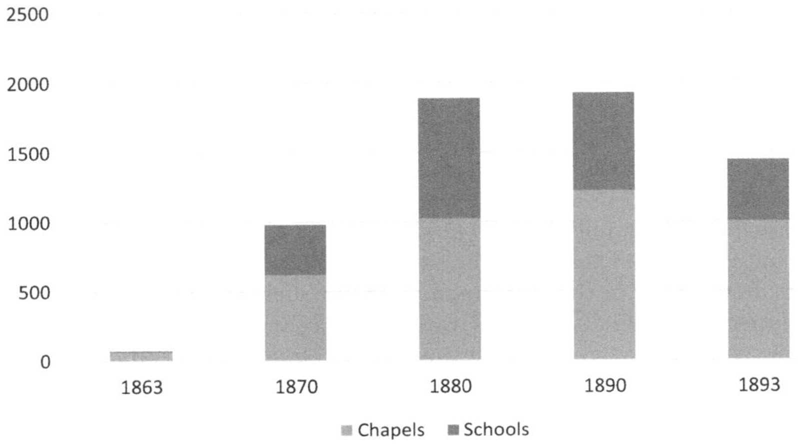

Chart 6.3 LMS chapels and schools, 1863–1893. (Source: Adapted from Campbell 1988b, 64)

chased material or skills. Brick churches, costing an average of $250–300 plus labour to erect, also became common in country districts. As sun-dried bricks could only be manufactured during the dry season, chapel and school construction occurred from April to November, when it impinged heavily either on the busiest agricultural seasons (harvest and preparation of the fields) or on the intervening period when most social ceremonies were held (LMS 1876, 33, 35; 1877, 27; 1890, 88; 1910, 10; Sibree 1880, 44). Moreover, as clay was generally obtained by digging pits close to rice fields, and then mixed with water from nearby rice fields, irrigation channels, streams or rivers, before being moulded and dried—a process sometimes interrupted by rainfall—and then fired, such activity created small pools of stagnant water that could provide mosquitoes with breeding sites (Sipa 2018; Grifa et al. 2017; Marrama 1995).

The state-church was also used to recruit military and industrial *fanompoana*. For example, from the close of 1879, when war with France appeared inevitable, the court relaxed its all-Merina rule for the army and started to enrol subject peoples, notably from Betsileo. The risks inherent in the abandonment of the traditional policy of keeping the imperial army "Merina" were more than outweighed by the fear of domestic revolt should the burden of *fanompoana* in Imerina continue to increase (LMS 1882, 1884). Conscription was achieved through compulsory registration in schools and churches. In 1882, the announcement of a 6000-man military conscription in Betsileo caused panic amongst church congregations there. However, missionaries generally welcomed the increase in school numbers. Lars Vig (1845–1913), the Norwegian Missionary Society (NMS) agent in Masinandraina district, even went as far as to successfully request the forcible registration of a further 1100 scholars in his schools. Similar demands were made by missionaries of other denominations. Missionaries also willingly complied with an imperial order that pupils regularly practise military drills, and during an invasion threat between May and November 1883, all schools and colleges devoted half of their time to spear and shield drill (Campbell 1988b, 59–60).

In 1884, the LMS had 73,324 scholars registered in Imerina, and a further 20,683 in Betsileo, and although the number of registered pupils fluctuated, they remained at about 100,000 until 1887. In January 1888, pastors were warned that if school numbers were not maintained, they too would be drafted into the army. Subsequently all boys over 16 years of age were called up. However, large numbers fled before they could be drafted. So great was the loss of scholars through conscription or flight that, by early 1889, most schools in the Southern Betsileo regions of Lalangina, Southern Isandra and Arindrano had collapsed, and in order to maintain the remainder, Protestant and Catholic teachers regularly raided each other's schools to "kidnap" pupils (Denjoy to Prime Minister 9 mai 1889, HH9, ANM; Borchgrevink to Prime Minister, 22 May 1889, HH6, ANM; Thorne to Wardlaw Thompson, 4 Jan 1888, FJKM; LMS, *Ten Years Review (1880–1890)*, 33, 93, 107; LMS 1876, 52–53; 1879, 4–6; 1884, 41). From 1890, as hope of British government support for their cause faded, the Merina court felt obliged to further augment its armed forces. For instance, there was a massive impressment of peasants into the army in Northern Betsileo (Berbezier to Rainisoavahia 16 juin 1891, HH9, ANM). In the early 1890s, as intimations grew of a French invasion, recruitment from schools became general. To protect their "gifted" pupils, the missionaries exposed to the military draft those pupils categorised as *vaventy sady malaina* ("old enough and lazy") (Haile to Prime Minister, 28 Aug 1889, HH1, ANM; Denjoy to Prime Minister, 9 mai 1889, HH9, ANM; SPG 1893, 787). Each year, some 40 per cent of pupils were ordered into the army, generally those the missions considered the dullest or most resilient to missionary teaching. For instance, the NMS missionary Peder Nilsen-Lund (1842–1914) sent 200 of his *malaina* pupils to the army in 1887, and a further 695 in 1892, in which year his missionary colleague in Ambatofinandrahana, 50 miles west of

Ambositra, despatched to the army 264 *malaina* out of a total of 695 pupils (Nilsen-Lund to PM, 9 Alohotsy 1892, Boks 270.F, FLM; Selmer to Rasoavahia, Fandriana, 23 Jul 1890, HH7, ANM; Nilsen-Lund to Prime Minister, Ambatofinandrahana, 5 Apr 1892, HH7, ANM).

From the 1880s, state-church schools were also used to recruit industrial labour. This was in part directly linked to the war effort, as when Rainimanana, Merina governor of Fihasinana, summoned school children to mine lead. However, from 1883, in order to obtain the means of paying the cost of the 1882–1885 conflict and the subsequent indemnity of $2,000,000 imposed on the Merina court by the French, most industrial *fanompoana* was used to exploit crown goldfields (Campbell 1988c). The chief crown mining areas were the river Onive, in the Ankaratra Mountains, the eastern forests bordering Betsileo, the Ampasay and Sakaleona basins and the Fisakana region. The region governed by Ambositra was particularly rich in alluvial gold and was heavily exploited, especially towards the close of the nineteenth century (Catat 1895, 178; Devred 1947, vol. 2, 64; Ravelomanana 1971, 21).

The failure to attract substantial foreign capital and skill meant that almost all gold was produced by slow and arduous labour-intensive methods. Work generally commenced in the dry season with the construction of a barrage across the river to create an expanse of tranquil water. The river bed was subsequently excavated, pits sometimes being sunk 3 metres, as gold deposits tended to be richer on the bedrock. There was a sexual division of labour with men digging while women removed the excavated soil in bateas (wooden pans) to the nearest water source to be washed. Nuggets several ounces in weight were occasionally found, and where veins were discovered, workers would prospect and exploit in every direction until all traces of ore vanished. During the rainy season, the men sometimes worked up to their shoulders in water, although (as that was the period when gold was being actively deposited) they did not need to dig deep. The same pits were frequently excavated following a flood or heavy rain, but the fresh layer of silt rarely yielded much gold (Marriot 1905, BRA; Letcher 1936, 273; Colville 1893, 205–206). Such activities created multiple pools of stagnant water, which proved ideal breeding ground for mosquitoes in gold-bearing areas linking lower-lying malaria-endemic regions to the traditionally malaria-free central highlands. Indeed, genetic studies indicate that *Anopheles funestus*, the main malaria vector in the highlands, spread there from lowland Eastern Madagascar—across the extensive gold-bearing zone straddling the eastern forest and highlands (see Map 6.3) (Ayala et al. 2006).

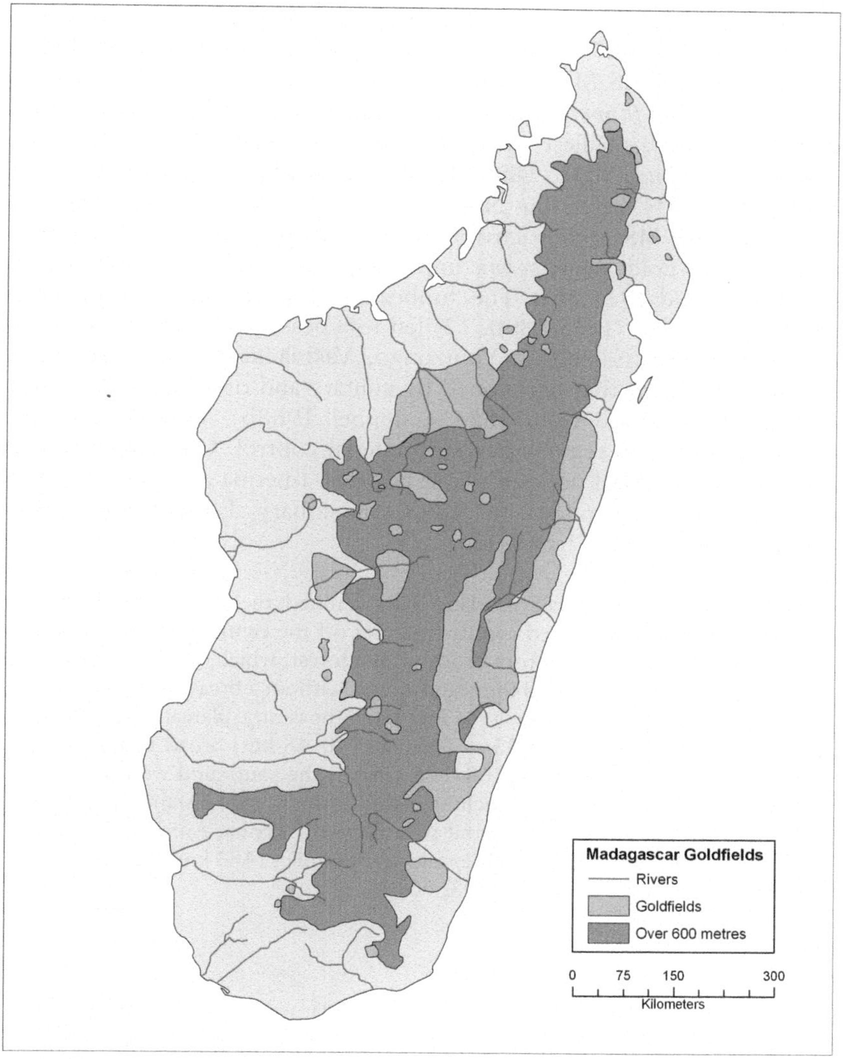

Map 6.3 Madagascar. Goldfields. (Source: Carl Hughes, IOWC)

Missionaries, as part of their contribution to the 1882–1885 war effort, organised "voluntary" school and church gold corvées ("Minute Book" 28 June 1884, CMS-MAD), while from 1885, the imperial court organised massive gold *fanompoana* that in Betsileo comprised "all the scholars, except the very smallest" irrespective of sex (LMS 1890, 33, 93; *Madagascar Times*, 8 Dec 1888; ANM, Haile to Prime Minister, 28 Aug 1889, HH1, ANM). In the Northern Betsileo regions of Mananadona and Fisakana, NMS church activities were interrupted by the imposition of a 400-strong gold *fanompoana* unit comprising both sexes as early as 1887 (Crosfield 1911, 96). The "unbearable labour connected with the working for gold" (LMS 1890, 12) led to a peasant flight from the land from the 1880s, notably in Vonizongo, Vakinankaratra and Northern Betsileo, similar to that occasioned by military and other fanompoana in Imerina in the period 1820–1854 (Campbell 1988b). Many, particularly young men, fled to regions beyond imperial control. Consequently, the *efitra*, or uninhabited no-man's land between Imerina and other regions of the island, expanded. In 1893, missionary Edward McMahon (1860–1918) stated (Chart 6.4):

> Very nearly if not quite half the island is uninhabited; there is a pretty clearly-defined belt of uninhabited country, cutting off the central provinces from the coast tribes; this comprises nearly all the forest, which, not always worthy of the name, runs round the island almost without a break, I believe, and also the rolling downs between the forest and the central plateau. This uninhabited part averages from 25 to 30 miles [40–48 km] broad to the east, and from 100 to 120 miles [160–193 km] on the south and west sides of Madagascar; the coast tribes reaching inland to an average distance of about 30 miles [48 km]; and the central provinces of Imerina, Antsihanaka and Betsileo being together, roughly speaking, 300 miles [483 km] long and 70 [113 km] broad. (McMahon 1893, 90)

He continued:

> Since the time of Radama I this region has enlarged from about twelve to twenty miles [19 to 32 km], and what was four or five generations ago, a populous and most fertile district, as the ruins of villages and the deserted rice-fields show, is now a barren wilderness, given over to wild-hogs and tall grass. I have crossed that part in different places, ranging from South Betsileo to the north-west of Imerina, and noticed the same thing was apparent all along the boundary line. (ibid., 92)

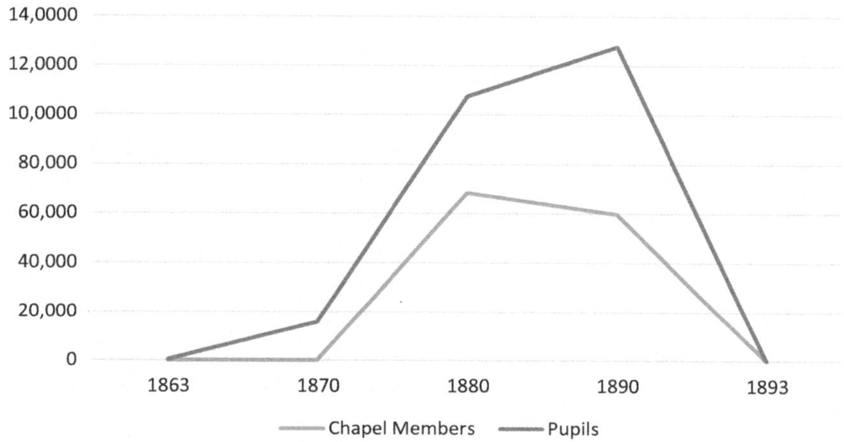

Chart 6.4 LMS chapel members and school pupils, 1863–1893. (Source: Adapted from Campbell 1988b, 64)

Most refugees from compulsory forced labour joined brigand bands, or independent rebel republics such as Ikongo, to the east of Fianarantsoa, which in 1853 boasted some 3000 warriors and in 1865 reputedly had a population of 30,000 (Raombana [1853b], 14, 17–18; Rooke 1866, 62). Such rebel groups engaged in raiding rural villages for cattle and slaves, deepening the inability of remaining villagers, often the old and infirm, to maintain the water channels necessary to a system of irrigated riziculture that in Betsileo was more complex than in Imerina, expanding up hillsides, in spectacular tiered rice terraces covering both convex and concave mountain slopes (Campbell 2005a, 24). The abandonment of rice fields and the complex system of canals that fed and drained them in turn offered the malarial vectors ideal breeding grounds in the Malagasy highlands (Charts 6.5 and 6.6).

There, as on mainland Africa, malaria became the major killer disease of all age groups. It was of particular demographic importance that a pregnant woman's acquired resistance to malaria decreases with the length of the pregnancy, heightening the risk of anaemia, which was a frequent cause of neonatal death. Surviving babies were, up to their fourth month, generally protected from the disease by foetal haemoglobin and antibodies from the mother. Hence malaria most seriously affected infants aged three to five, notably between the end of the rainy and the start of the winter season, when their physiological defences were often weakened by nutri-

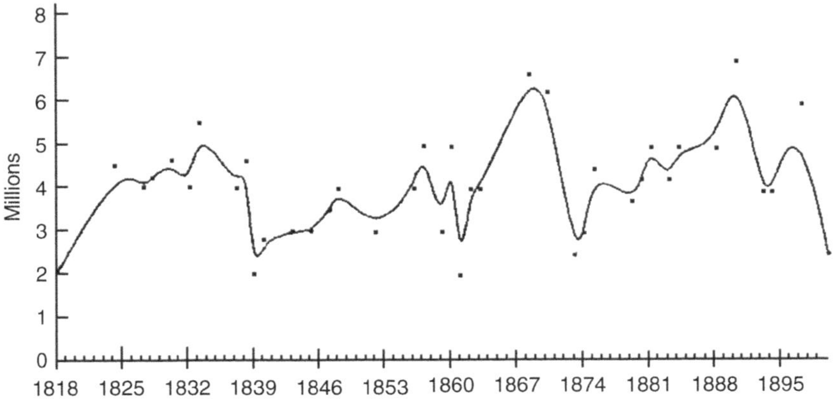

Chart 6.5 Madagascar: Population estimates. (Source: Campbell 2005a, 137)

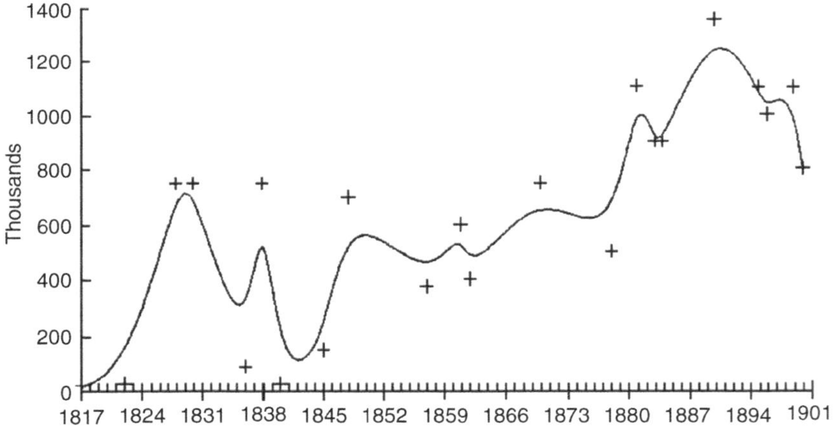

Chart 6.6 Imerina: Population estimates. (Source: Campbell 2005a, 139)

tional deficiency. In the early 1890s, McMahon considered that 75 per cent of infants died before the age of three (McMahon 1893, 93). However, it could seriously affect all age groups, especially when several years of poor rains and low malarial incidence, reducing the body's natural defences against infection, were followed by a protracted rainy season (Delmont 1982; Paillard 1987).

There occurred particularly severe outbreaks of malaria in Vonizongo, in 1874 (Matthews 1904, 100), and across the entire central plateau from 1877 to 1880, 1890 and 1895–1896. During the extended 1877–1880 epidemic, LMS missionary Thomas Matthews (1842–1928) estimated that in the 1878 outbreak "over the central provinces about forty per cent of those who were seized died", the impact being "exceptionally severe" in Vonizongo (ibid., 221). Matthews described the 1879 outbreak in Imerina as "much more severe than that of the previous year. The former had probably carried off 5000 people, yet it was mild when compared with what we had in 1879" (ibid., 208). In the capital of Antananarivo, NMS medical missionary, Carl Güldberg (1846–1901), saw 111 cases of malaria in 1877, 342 in 1878 and 889 in 1879 (Davidson 1895, 334). The malaria epidemic appears to have had a considerably greater impact in Betsileo than Imerina: in the 1878 outbreak, it killed an estimated 10,000 people in Southern Betsileo alone (LMS 1878, 57; see also ibid., 75, 106; Table 6.1). By contrast, from 1883 to 1886, Güldberg encountered only 93 cases of malaria (Davidson 1895, 334). This may have reflected less a dramatic subsidence of malaria than a huge rejection by the Malagasy of the medicine and religion of Westerners, who they widely believed to be the cause of the disease, and a return to traditional religious beliefs and methods of countering the disease (LMS 1878, 41; 1879, 26, 33, 47, 70; Chapus 1925, 241–242). Thus the missionaries observed in 1878:

> Nearly all the churches and schools were broken up and in many instances the terrified people sought out the old diviners and soothsayers, trusting to escape the disease through their charms and incantations. (LMS 1878, 33; see also LMS 1879, 8)

The region around Fianarantsoa, in Betsileo, experienced in May 1890 "a very severe epidemic of remittent fever" (Richard Baron quoted in Moss 1913, 142). The widespread malaria epidemic of 1895 caused 25 per cent mortality (Campbell 1991, 433). The following year it again caused high mortality in Southern Betsileo (Moss 1913, 164–165). By 1905, the only area of the plateau free from malaria was Antsirabe, a town lying roughly halfway between Antananarivo and Fianarantsoa (Campbell 1991, 432–433). As a result, not only did ever increasing numbers of Malagasy abandon the Christian for the traditional medical system, many also joined the Menalamba Revolt that exploded in 1895 and which was primarily directed against the missionaries, indigenous church officials and members of the Merina state that protected them (Campbell 1988b) (Map 6.4).

Summary

European accounts from the sixteenth century attest to endemic malaria on the coasts and hinterland lowlands of Madagascar. In the late eighteenth century, when Europeans first ventured into Imerina, in the highland interior, they noted it to be malaria-free. However, by 1905 the only area of the highland plateau free from malaria was in Antsirabe, a town lying roughly halfway between Antananarivo and Fianarantsoa. Some historians have indicated that the first outbreak of malaria on the plateau occurred in 1878. However, closer analysis reveals not only that malaria was endemic in some highland zones at least by the early nineteenth century but that malaria outbreaks occurred on the plateau in epidemic form probably as early as the 1820s. These outbreaks were connected both to changing climatic conditions, notably warmer weather that facilitated the survival of mosquitos at higher altitudes, and to human policies, particularly to *fanompoana*, or unremunerated forced labour for the Merina state. From 1820 to 1853, *fanompoana* led to unprecedented flows of people between areas where malaria was endemic and the traditionally malaria-free regions of the highlands. It also resulted in high mortality for Merina subjects and flight from the land—with the consequence that many rural areas became depopulated, and rice fields deserted, creating networks of stagnant water that were potentially ideal breeding grounds for mosquitoes.

In 1869, the Merina court created a state-church and school system that not only required large amounts of labour to create and maintain but also formed the basis for the registration of forced labour recruitment for the state. State *fanompoana* for males was channelled primarily into the army which faced mounting threats from French forces, and for men, women and children into goldfields. Moreover, to avoid the possibility of revolt in Imerina, forced labour was concentrated in both Imerina and Betsileo. It proved so exploitative that many small cultivators fled, abandoning highly intricate irrigated agriculture systems—which consequently formed, alongside abandoned alluvial gold diggings and clay-brick pits, networks of stagnant water ideal for the *Anopheles funestus* vector of malaria to spread from the eastern lowlands into the highlands.

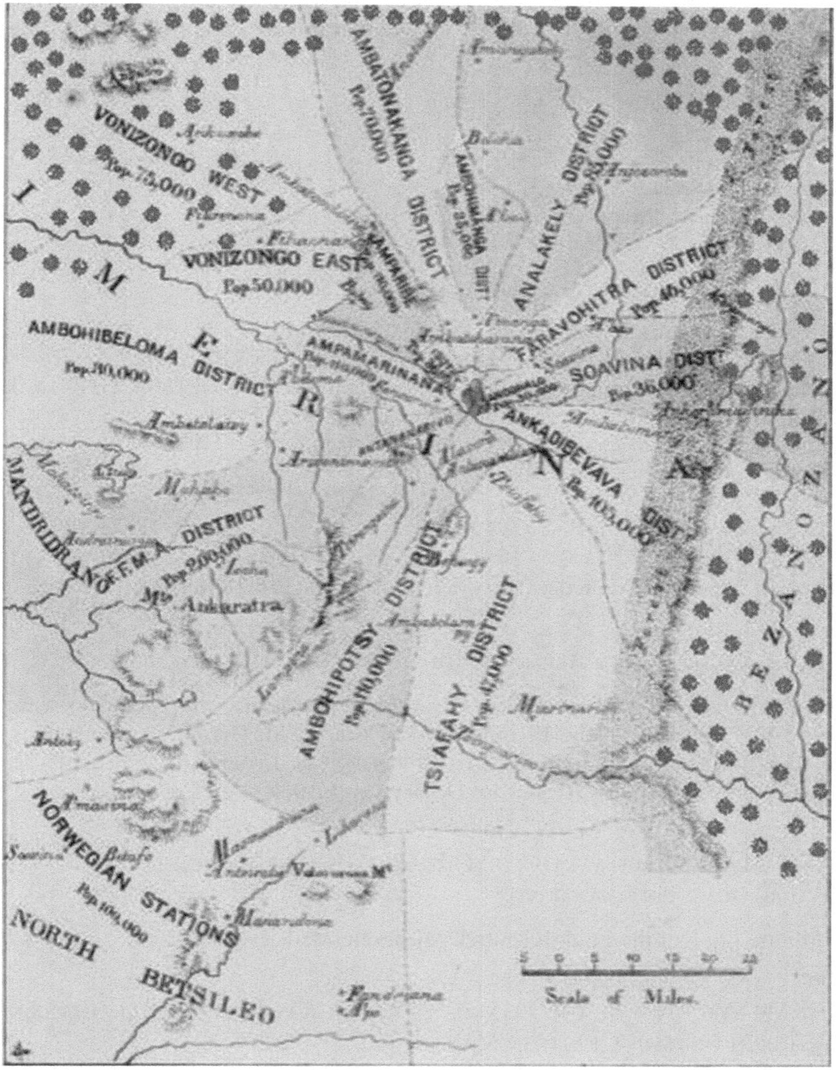

Map 6.4 Imerina: districts and endemic malarial zones (marked in red). (Source: Adapted from LMS 1890, 18–19)

References

Archival Sources

AAM—Archives de l'Académie Malgache

Raombana. 1853a. "Annales".
——— [1853b] "Manuscrit".

ANM—Archives Nationales de Madagascar

Berbezier to Rainisoavahia, Betafo, 16 juin 1891, HH9.
Borchgrevink to Prime Minister, Antsahamanitra, 22 May 1889, HH6.
Denjoy to Prime Minister, "private", Ambodinandohalo, 9 mai 1889, HH9.
Haile to Prime Minister, Ambohibeloma, 28 Aug 1889, HH1.
Lund to Prime Minister, Ambatofinandrahana, 5 Apr 1892, HH7.
Selmer to Rasoavahia, Fandriana, 23 Jul 1890, HH7.

BL—British Library

Hugon, "Aperçu de mon dernier voyage à ancova de l'an 1808", Add.18137.

BRA—Barlow Rand Archives, Johannesburg, South Africa

Hugh Marriot, "Report on the Gold Resources of Madagascar with special reference to the properties of the Lecompte Madagascar Gold, Harming's Concessions, Harrison Smith's Concessions, Cie. Coloniale, Cie, Lyonnaise" (Johannesburg, 22 September 1905), HE Record Department F.10, No. 9, "Private Reports".

CMS-MAD—Archives of the Church Missionary Society in Madagascar, Antananarivo

"Minute Book of the English Church Mission in Madagascar".

FJKM—Archives of the *Fiangonan'i Jesoa Kristy eto Madagasikara* (Church of Jesus Christ in Madagascar)

Thorne to Wardlaw Thompson, 4 Jan 1888, Imerina District Committee Letter Book (1875–1891).

FLM—Archives of the *Fiangonana Loterana Madagaskara* (Lutheran Church in Madagascar)

Nilsen Lund to PM, Ambatofinandrahana, 9 Alohotsy 1892, Boks 270.F.

NAK—National Archives, Kew

James Hastie, "Diary" (1822), CO.167, vol. 63.
Le Sage, "Mission to Madagascar" (1816), CO.167/34.

NAM—National Archives of Mauritius

Barnsley, diary (1821–1822), HB 7.
C.M. Campbell, Journal (1840), HB 14.
John Joseph Freeman to Dick, Antananarivo, 13 Oct 1834, HB 9.
James Hastie to Telfair, Tananarive, 8 Oct 1821, HB 21.
——— to Barry, Port Louis, 18 Apr 1822a, HB 7.
——— to Farquhar, Foul Point, 14 Oct 1822b, HB 7.
——— to Cole, Marouvoie, 14 Aug 1824a, HB 5.
——— to [Barry], Marouvoie, 14 Aug 1824b, HB 5.
——— to Lowry Cole, Tananarive, 10 Dec 1824c, HB 5.
John Marshall, Report of the Voyage of Her Majesty's Ship *Isis* to the Comoro Islands, Mozambique Coast and Madagascar, 24th March 1843, HB 14.
Robert Lyall, Journal (1830), HB 20.

Other References

Anderson, Thomas. 2017. Converting the Hospital: British Missionaries and Medicine in Nineteenth-Century Madagascar. *Itinerario* 41 (3): 539–554.

Ayala, Diego, et al. 2006. Population Structure of the Malaria Vector Anopheles Funestus (Diptera: Culicidae) in Madagascar and Comoros. *Acta Tropica* 97 (3): 292–300.

Baron, Richard. 1878. Jottings on Some Plants of Imerina. *Antananarivo Annual and Madagascar Magazine* 4: 106–115.

Blanchard, R. 1907. Le paludisme à Madagascar. *Archives de Parasitologie* 11 (2): 185–214.

Campbell, Gwyn. 1987. The Adoption of Autarky in Imperial Madagascar, 1820–1835. *Journal of African History* 28 (3): 395–411.

———. 1988a. Slavery and Fanompoana: The Structure of Forced Labour in Imerina (Madagascar), 1790–1861. *Journal of African History* 29 (3): 463–486.

———. 1988b. Missionaries, Fanompoana and the Menalamba Revolt in Late Nineteenth Century Madagascar. *Journal of Southern African Studies* 15 (1): 54–73.

———. 1988c. Gold Mining and the French Takeover of Madagascar, 1883–1914. *African Economic History* 17: 1–28.

———. 1991. The State and Pre-colonial Demographic History: The Case of Nineteenth Century Madagascar. *Journal of African History* 32 (3): 415–445.

———. 1992. "Crisis of Faith and Colonial Conquest. The Impact of Famine and Disease in late Nineteenth-Century Madagascar" Cahiers d'Études Africaines 32 (3) 127: 409–453.

———. 2005a. *An Economic History of Imperial Madagascar, 1750–1895: The Rise and Fall of an Island Empire*. Cambridge: Cambridge University Press.

———. 2005b. The Debate over Malagasy Origins. *ZIFF Journal* 2: 5–14.

———. 2012. *David Griffiths and the Missionary 'History of Madagascar'*. Leiden: Brill.

———. 2018a. Introduction: Bondage and the Environment in the Indian Ocean World. In *Bondage and the Environment in the Indian Ocean World*, ed. Gwyn Campbell, 1–32. Basingstoke, Hampshire: Palgrave Macmillan.

———. 2018b. Environment and Enslavement in Highland Madagascar, 1500–1750: The Case for the Swahili Slave Export Trade Reassessed. In *Bondage and the Environment in the Indian Ocean World*, ed. Gwyn Campbell, 49–78. Basingstoke, Hampshire: Palgrave Macmillan.

———. 2019. *Africa and the Indian Ocean World from Early Times to 1900*. Cambridge: Cambridge University Press.

Carayon, Jean-Louis. 1845. *Histoire de l'établissement français de Madagascar pendant la Restauration précédée d'une description de cette ile et suivie de quelques considérations politiques et commerciales sur l'expédition et la colonisation de Madagascar*. Paris: Gide.

Catat, Louis. 1895. *Voyage à Madagascar*. Paris: l'Administration de l'Univers Illustré.

Chapus, George-Sully. 1925. *Quatre-vingts annees d'influence europeennes en Imerina (1815–1895)*. Tananarive: G. Pitot.

Chapus, George-Sully, and Gustave Mondain. 1954. *Le Journal de Robert Lyall*. Tananarive: Imprimerie Officielle.

Cohen, William B. 1983. Malaria and French Imperialism. *Journal of African History* 24 (1): 23–36.

Colville, Zélie. 1893. *Around the Black Man's Garden*. Edinburgh: Blackwood.

Coppalle, André. c.1827. Voyage dans l'intérieur de Madagascar et à la capitale du roi Radame pendant les années 1825 et 1826. Accessed 15/10/07. http://www.bextes.org/coppalle.pdf.

Cousins, William. 1895. Madagascar in the Year 1840. *Antananarivo Annual and Madagascar Magazine* 19: 341–344.

Crosfield, Albert, and Gulielma. 1911. *A Man in Shining Armour: The Story of the Life of William Wilson, Missionary in Madagascar*. London: Headley.

Davidson, A. 1895. The Diseases Prevalent in Madagascar. *Antananarivo Annual and Madagascar Magazine* 19: 330–337.

Delmont, Jean. 1982. Paludisme et variations climatiques saisonnières en savane soudanienne d'Afrique de l'Ouest. *Cahiers d'études africaines* 22 (85–86): 117–134.

Deschamps, Hubert. 1959. *Les Migrations intérieures passées et présentes à Madagascar*. Paris: Berger-Levrault.

Devred, E. 1947. Les mines. In *Madagascar*, ed. Marcel de Coppet, vol. 2, 63–85. Paris: Encyclopédie De l'Empire Français.

Ellis, William. 1838. *History of Madagascar. Comprising also the Progress of the Christian Mission Established in 1818; and an Authentic Account of the Recent Martyrdom of Rafaravavy; and of the Persecution of the Native Christians*. 2 vols. London: Fisher.

Freeman, Joseph John, and David Johns. 1840. *A Narrative of the Persecution of the Christians in Madagascar with Details of the Escape of the Six Christian Refugees Now in England*. London: John Snow.

Gaüzère, B.-A., and P. Aubry. 2013. Histoire des épidémies et des endémoépidémies humaines dans le sud-ouest de l'océan Indien. *Médecine et Santé Tropicales* 23 (2): 145–157.

Grandidier, Alfred & Guillaume. 1928. Histoire *Physique, Naturelle et Politique de Madagascar. vol. 4: Ethnographie de Madagascar*. Tome 4. Paris: Hachette et Société d'Editions Géographiques, Maritimes et Coloniales.

Grifa, Celestino, et al. 2017. Traditional Brick Productions in Madagascar: From Raw Material Processing to Firing Technology. *Applied Clay Science* 150 (15): 252–266.

Griffiths, David. 1843. *Hanes Madagascar, neu Grynodeb o Hanes yr Ynys, ei Chynyrch, ei Masnach, ac Ansawdd ei Thrigolion; yn nghyda'u Harferiadau Creulon, a'u Heilunaddoliaeth Ffiaidd*. Machynlleth: Richard Jones.

Guillain, Charles. 1845. *Documents sur l'Histoire, La Géographie et le Commerce de la Partie Occidentale de Madagascar*. Paris: Imprimerie Royale.

Hay, Simon I., et al. 2000. Etiology of Interepidemic Periods of Mosquito-Borne Disease. *Proceedings of the National Academy of Sciences* 97 (16): 9335–9339.

Hewitt, R., et al. 1996. Beta-Globin Haplotype Analysis Suggests that a Major Source of Malagasy Ancestry Is Derived from Bantu-Speaking Negroids. *American Journal of Human Genetics* 58 (6): 1303–1308.

Hlongwana, Khumbulani W., et al. 2009. Community Knowledge, Attitudes and Practices (KAP) on Malaria in Swaziland: A Country Earmarked for Malaria Elimination. *Malaria Journal* 8: 29. https://doi.org/10.1186/1475-2875-8-29.

IOWC Database of Human-Environment Interaction in the IOW.

Jennings, Eric. 2006. *Curing the Colonizers: Hydrotherapy, Climatology, and French Colonial Spas*. Durham, NC: Duke University Press.

Laverdant, Désiré. 1844. *Colonisation de Madagascar*. Paris: La Société Maritime.

Letcher, Owen. 1936. *The Gold Mines of Southern Africa: The History, Technology and Statistics of the Gold Industry*. Johannesburg: Letcher.

Lewis, Locke. 1835. An Account of the Ovahs, a Race of People Residing in the Interior of Madagascar: With a Sketch of Their Country, Appearance, Dress, Language, &c. *Journal of the Royal Geographical Society* 5: 230–242.
LMS. 1910. *Ten Years Review 1901–1910*. Antananarivo: LMS.
———. 1890. *Ten Years Review 1880–1890*. Antananarivo: LMS.
———. 1884. *Report of the Madagascar Mission for 1884*. Antananarivo: LMS.
———. 1882. *Report of the Madagascar Mission for 1882*. Antananarivo: LMS.
———. 1879. *Report of the Madagascar Mission for 1879*. Antananarivo: LMS.
———. 1878. *Report of the Madagascar Mission for 1878*. Antananarivo: LMS.
———. 1877. *Report of the Madagascar Mission for 1877*. Antananarivo: LMS.
———. 1876. *Report of the Madagascar Mission for July 1 1875–June 30 1876*. Antananarivo: LMS.
Mackay, James G. 1893. *Some Notes on Native Medicine* and Medical Customs, as Practiced by the Sihanaka. *Antananarivo Annual and Madagascar Magazine* 17: 45–54.
McMahon, E.O. 1893. The Population of Madagascar: Does It Increase or Decrease? *Antananarivo Annual and Madagascar Magazine* 17: 90–94.
Marrama, L. 1995. Anopheles Funestus and Rice Agriculture in the Madagascar Highlands. *Sante* 5 (6): 415–419.
Matthews, Thomas Trotter. 1904. *Thirty Years in Madagascar*. London: Religious Tract Society.
Milne, William. 1974 [originally published 1815]. Island of Madagascar. In *Travels in South Africa*, ed. John Campbell, Appendix 5. Cape Town: C. Struik.
Moss, C.F.A. 1913. *A Pioneer in Madagascar. Joseph Pearse of the LMS*. London: Headley Brothers.
———. 1895. The Diseases Prevalent in Madagascar. *Antananarivo Annual and Madagascar Magazine* 19: 330–341.
Mouchet, J., et al. 1997. La reconquête des Hautes Terres de Madagascar par le paludisme. *Bulletin de la Société de Pathologie Exotique* 90 (3): 162–168.
Oliver, Samuel Pasfield. 1886. *Madagascar. An Historical and Descriptive Account of the Island and Its Former Dependencies*. 2 vols. London: Macmillan.
Paillard, Yvan-Georges. 1987. Les recherches démographiques sur Madagascar au début de l'époque coloniale et les documents de 'l'AMI'. *Cahiers d'Études africaines* 27 (105–106): 17–42.
Parker, G.W. 1881. A Malagasy Materia Medica. *Antananarivo Annual and Madagascar Magazine* 5: 61–63.
Pearse, Joseph. 1884. Medical Missionary Work in Madagascar by a Non-Professional. *Antananarivo Annual and Madagascar Magazine* 8: 11–27.
Pfeiffer, Ida. 1881. *Voyage à Madagascar [1857]*. Paris: Librairie Hachette.
Radaody-Ralarosy, P. 1971. Proverbes et pathologie. *Bulletin de l'Académie Malgache* 49 (2): 45–48.
Raison-Jourde, Françoise. 1991. *Bible et pouvoir à Madagascar au xixe siècle: Invention d'une identité chrétienne et construction de l'État*. Paris: Karthala.

Rakotoson, Jean-Desire, et al. 2017. Insecticide Resistance Status of Three Malaria Vectors, Anopheles Gambiae (s.l.), An. Funestus and An. Mascarensis, from the South, Central and East Coasts of Madagascar. *Parasites & Vectors* 10: 396. https://doi.org/10.1186/s13071-017-2336.

Randrianarivelojosia, Milijaona et al. 2003. "Plants Traditionally Prescribed to Treat *Tazo* (Malaria) in the Eastern Region of Madagascar" *Malaria Journal*, 2.25. Accessed 12/08/10. http://www.malariajournal.com/content/2/1/25.

Rasoanaivo, Philippe, et al. 1992. Medicinal Plants Used to Treat Malaria in Madagascar. *Journal of Ethnopharmacology* 37 (2): 117–127.

Razafimahatratra, Raymond. 2005. *The Sustaining Power of the Bible to the Martyrs During the Persecution in Madagascar from 1828 to 1861: Historical and Hermeneutical Analysis.* Master of Theology thesis, School of Religion and Theology, University of Kwazulu-Natal, Pietermaritzburg.

Robequain, Charles. 1958. *Madagascar et les bases dispersées de l'union française.* Paris: Presses universitaires de France.

Rooke, W. 1866. A Boat-Voyage Along the Coast-Lakes of East Madagascar. *Journal of the Royal Geographical Society* 36: 52–64.

Sibree, James. 1870. *Madagascar and Its People. Notes of a Four Year's Residence.* London: Religious Tract Society.

———. 1880. *The Great African Island: Chapters on Madagascar; A Popular Account of Recent Researches in the Physical Geography, Geology, and Exploration of the Country, and Its Natural History and Botany.* London: Trubner & Co.

———. 1915. *A Naturalist in Madagascar.* London: Seeley.

SPG. 1893. *Society for the Propagation of the Gospel in Foreign Parts, 1701–1892.* London: SPG.

Thorne, James Cross. 1888. Elementary Education in Madagascar. *Antananarivo Annual and Madagascar Magazine* 9: 27–40.

Theses

Ravelomanana, Jaqueline. 1971. *La vie religieuse à Ambositra, 1880–1895.* Thèse, Université de Madagascar.

Newspapers

The Madagascar Times (1888).

Internet References

Miller, Katie. 2018. *Plasmodium Life Cycle.* Accessed 15/09/18. http://www.intellectualventureslab.com/invent/plasmodium-life-cycle.

Sipa, Masika. 2018. A Life for Bricks. *MadaMagazine.* Accessed 15/09/18. https://www.madamagazine.com/en/ein-leben-fuer-backsteine/.

CHAPTER 7

Disease, Alcohol Consumption, and Excise in Nineteenth-Century British India

Peter Hynd

Introduction: Death and Taxes

1897–1898 was a good financial year for excise revenues in the Bombay Presidency. Revenue from sales of distilled liquor alone had risen by a staggering 326,000 rupees—an increase of more than 5 per cent compared to 1896–1897 (Bombay Excise Report 1897–1898, 4). More than half of this increase came from the city of Bombay and the neighbouring Thane District. Given that the Abkari (Excise) Department was a subsidiary of the provincial Board of Revenue, one might expect such results to have been cause for official celebration. Instead, the reaction of the Commissioner of Customs, Salt, Opium, and Abkari was guarded. Although keen to cite modest increases in liquor revenue as evidence of improved administration, top excise bureaucrats were well aware that the increasingly vocal critics of British India's excise policy—temperance activists, missionaries, Indian nationalists—were sure to see the dramatic increase in liquor revenues as further proof of their claims that the Government of India was

P. Hynd (✉)
McGill University, Montreal, QC, Canada
e-mail: peter.hynd@mcgill.ca

G. Campbell, E.-M. Knoll (eds.), *Disease Dispersion and Impact in the Indian Ocean World*, Palgrave Series in Indian Ocean World Studies, https://doi.org/10.1007/978-3-030-36264-5_7

deliberately fostering insobriety for the sake of generating excise revenue. No doubt aware that these numbers would invite criticism, John Pollen (1855–1923), the Acting Commissioner of Customs, Salt, Opium, and Abkari in Bombay, offered a simple justification in his annual report to the provincial government:

> The increase in Bombay is chiefly due to the fact that people, having become familiar with the plague, did not leave [the city] on its outbreak during 1897–98 in such large numbers as the previous year. There was consequently a larger liquor-consuming public in the town, but besides this the wide-spread belief that liquor constitutes somewhat of a prophylactic against plague, probably led to increased consumption. (Bombay Excise Report 1897–1898, 4)

Taken at face value, the return to Bombay of people who had fled the plague in 1896–1897 and the spread of a belief that liquor offered some protection against infection were a perfectly reasonable explanation for the spike in excise revenues in 1897–1898. Records from 1896–1897 showed a decline of 119,117 rupees in excise revenue for the city of Bombay alone. This was, at the time, attributed 'to the decreased consumption of country liquor owing mainly to the [initial outbreak of] plague' which had caused panic, disruption, and severe mortality when it had first hit the city in the summer and autumn of 1896 (Bombay Excise Report 1896–1897, 6). People remaining in the city after having fled during the worst days of the initial outbreak the previous year could easily explain the increased excise revenues of 1897–1898, even without taking into account a popular belief in the prophylactic power of potent liquor. Although sales of liquor were up for 1897–1898, excise revenues remained lower than they had been in the financial year before the outbreak of plague. The Commissioner's explanation was completely plausible.

However, a deeper look at references to plague and other diseases in the records of the excise administration complicates the picture. References to plague are numerous, but the cited effect on excise revenues differs wildly from case to case. For example, the 1897–1898 Bombay excise report cited above—the same that attributed the *increase* in the excise revenue of Bombay and Thane Districts to local responses to the plague—placed the blame for a substantial *decline* in the excise revenues of Bharwar and Buldana on the enduring presence of plague in those districts (Bombay Excise Report 1897–1898, 5). The excise report for 1898–1899 noted that the Government of Bombay had been forced to grant concessions to liquor contractors who had paid it considerable sums in exchange for three-year, district-wide monopolies, 'because famine and plague restrictions had, in a manner altogether beyond [their] control, adversely affected sales in [their] districts' (Bombay Excise Report 1887–1889, 4). Nevertheless, and without any apparent sense of irony, the very next paragraph cheerfully attributed the large increase in liquor sales in some of those same districts to the relaxation of plague restrictions and explained the continued growth of Bombay's excise receipts—which by then had surpassed pre-plague levels—as the result of 'larger consumption of liquor both among labouring classes and among classes which generally abstain from drink, as the popular impression has grown that liquors act as a tonic or preventative against plague' (Bombay Excise Report 1887–1889, 4).

Such claims, despite their sometimes contradictory nature, and the lack of clear evidence for a correlation between plague and excise revenues, may have been reported in good faith. However, what emerges from a careful study of these and other references to the effects of plague and other diseases on excise revenues is a profound disinterest on the part of excise officials in the underlying truth of such claims. Late nineteenth- and early twentieth-century excise reports from across British India make repeated reference to liquor being used to prevent, treat, or mitigate the effects of deadly diseases such as cholera and the bubonic plague. Nevertheless, official reports linking outbreaks of epidemic disease to both exceptional growth and dramatic decline in excise revenues were common. If any serious studies were undertaken to understand how Indians made use of liquor as a medicine, or to prove or disprove any of the alleged links between disease outbreak and excise revenue, they do not appear to have attracted any attention within the Abkari Department. Given this official disinterest, the fact that excise officials repeatedly invoked disease to explain fluctuations in excise revenues is extremely telling. This chapter will examine official claims about the relationship between disease, alcohol consumption, and excise revenue in late nineteenth-century British India and evaluate them in light of contemporary statistical evidence regarding disease mortality.

Background: Excise and the Raj

Throughout British India the Abkari Departments' interest in disease stemmed from the fact that provincial excise officials—namely, the provincial Commissioners of Excise and the district-level Collectors of Revenue—

felt compelled to justify any sudden increase or decrease in excise revenues (Hynd 2020).[1] Rapid growth in excise revenues might please the provincial Board of Revenue, but from about the mid-nineteenth century it also agitated the excise administration's increasingly vocal critics: missionaries, temperance advocates, and nationalists. Dramatic year-to-year increases attracted unwelcome outside attention, and unexplained figures showing growth in the range of thousands or tens of thousands of rupees within a single year offered opponents an easy avenue of attack. Sudden declines in excise revenues, on the other hand, often drew the ire of superiors within the revenue-hungry colonial government and, if interpreted as poor performance or neglect of duty, could damage a young officer's future career prospects. For the purposes of mounting a defence against external critics or justifying poor results to superiors, claims that the people were turning to liquor as defence against plague or cholera were sufficient.

The history of excise officials in British India haphazardly citing disease as the cause of an increase or decrease in excise revenues, or making other unsubstantiated claims about the medicinal uses of alcohol, can be traced to at least the 1840s. Such references became increasingly common during the second half of the nineteenth century, as the provincial excise establishments of British India steadily expanded their reach and ability to extract revenue from alcohol sales. This expansion, coupled with the growth of an international temperance movement, generated ferocious criticism of the Government of India's excise policy both locally and abroad. The bureaucrats who managed the collection of excise taxes found themselves caught between the provincial governments' ravenous appetite for revenue and a need to justify the growth of such revenues in the face of mounting external criticism. In this environment, disease, as something that could plausibly impact excise revenues, but remained entirely beyond the control of the excise bureaucracy, was a perfect excuse when policies and regulations failed to produce the desired outcomes. Throughout the period of rapid growth of the excise administration in British India, from approximately the middle of the nineteenth century to after the First World War, excise officials frequently invoked the presence of disease to

[1] More detail on excise in British India can be found in my forthcoming PhD thesis 'Excise and Alcohol in British India' (McGill University, 2020). The information contained in the following paragraphs is derived from my thesis work, which was based on the detailed study of the official archives of the excise departments of Bengal, Bombay, and Madras from the foundation of these departments until approximately 1920.

account for and justify both sudden increases in excise revenue—which would anger their critics—and sudden decreases, which would otherwise appear to result from poor job performance. Major disease epidemics were especially useful for this purpose because they were sudden and unpredictable. They could easily and plausibly explain away dramatic year-to-year fluctuations and other large statistical anomalies, but, as 'Acts of God', never required the usually ponderous excise bureaucracy to make dramatic adjustments to carefully calculated rates of duty and other regulations.

Consequently, disease appeared far more than one might initially expect in excise department reports in late nineteenth-century British India. It always did so as an unpredictable variable, cited to provide justification for unexpected results. Such references were intended to be taken at face value, without further examination. Excise bureaucrats presented disease as a factor that could not be accounted for. The impact of disease upon excise revenues therefore neither warranted detailed investigation nor necessitated major changes to existing regulations. Superficially, this was reasonable; short of preventing adulterated or exceptionally potent alcohol from reaching consumers, provincial excise departments were not responsible for public health. It was beyond their means to predict disease outbreaks, and they had neither the mandate nor the ability to influence public perceptions regarding the medicinal value of alcohol or lack thereof. Few official references to the impact of disease on excise revenue were ever challenged.

A brief overview of excise and alcohol regulations in British India is necessary before proceeding to a detailed investigation of the way excise bureaucrats used disease to explain and defend their annual results. The East India Company first began collecting excise taxes on alcohol production in the late eighteenth century. Initially, this resembled a system of revenue farming. Monopoly rights to produce and sell liquor in a given district were sold, often informally, to the highest bidder, who was then free to conduct his trade with little further interference from the Company. The first comprehensive excise regulations in India were penned in the Bengal Presidency in 1792, and soon after transplanted, almost word for word, to Bombay and later Madras. These early regulations formalised the system for licensing the production and retail sale of liquor. Where population density permitted, the law allowed for the concentration of liquor production, but elsewhere unsophisticated systems of monopoly renting and revenue farming remained the principal method by which the colonial state raised excise revenue. From the 1840s, however, the provincial governments of British India took increasingly direct control over excise taxation. By the 1870s, most provinces boasted sophisticated excise departments, the largest of which employed hundreds of bureaucrats, inspectors, and assistants. Where they could be enforced, regulations were gradually introduced that required shopkeepers to hold (costly) licences, and liquor manufacture was gradually confined to defined areas where excise department agents could guarantee that each gallon of liquor distilled would pay a set rate of duty before passing out of the distillery and into the hands of the retailer. These regulations, combined with vigorous efforts to stamp out unlicensed (therefore, illicit) home distilling, were remarkably effective, and excise revenues derived from liquor grew rapidly. Negligible at the beginning of the nineteenth century, excise taxes on alcohol, primarily distilled liquor, accounted for nearly 10 per cent of all government revenues in India by 1900.

Small in number but extremely vocal, temperance activists attacked the excise policy of the Government of India on both religious and secular grounds. On more than one occasion in the 1870s and 1880s, members of the Opposition with temperance sympathies openly attacked the Government of India's excise policies in the House of Commons, thoroughly embarrassing the sitting Secretary of State for India. From within India, protests, petitions, and boycotts against individual liquor merchants become commonplace from the late 1870s. By the 1890s, the Government of India had defaulted to blaming such campaigns on nationalist 'agitators', despite plenty of evidence to suggest that until the beginning of organised, nationalist-led campaigns during the First World War, many if not most of these protests were local affairs triggered by economic, social, caste, or religious concerns. The diverse cast of critics each had their own reasons for opposing the excise administration, but all agreed that the rapid growth of excise revenues was clear evidence of the colonial state actively promoting liquor consumption, with dire consequences for the moral, spiritual, and financial well-being of India's poorest people.

Government officials, particularly excise administration agents, loudly rejected this interpretation. They argued that growing excise revenues resulted directly from an improved excise administration that was gradually bringing all liquor production under government supervision, where it could be regulated and taxed for the public good. The official guiding maxim of excise policy, its defenders were keen to note, was 'Maximum Revenue from Minimum Consumption', a theory that identified taxation as the principal mechanism by which the state could control and limit access to destructive and deleterious liquor. According to this maxim, there was no way to eliminate the consumption of alcohol entirely. However, by inflating the retail price of liquor through taxation, the state could minimise the negative social effects of alcohol consumption by limiting the quantity of alcohol consumers were able to purchase. The majority of consumers of domestically produced 'country liquor'—the largest source of excise revenue in most provinces—were assumed to be poor people with extremely limited disposable income, most or all of which they would spend on liquor or other drugs, if given the chance. It fell to the Commissioner of Excise, in consultation with local collectors and their excise deputies, to determine the exact 'limit of taxation' in a given town or district. Setting excise taxes too low would result in the retail price of liquor falling and, according to the prevailing theory, lead to a rise in alcohol consumption (and associated menace of public intoxication) without any increase in excise revenue. Setting excise taxes too high, however, would tempt habitual consumers towards illicit sources of liquor, lowering excise revenues and promoting criminal activity without ever reducing the base level of consumption. The perfect rate of taxation, accompanied by efforts to suppress illicit distillation, would, it was argued, result in a retail price that was low enough to permit habitual consumers of liquor to satisfy their cravings from licit sources, but not so low that it became easy to overindulge.

This was a powerful argument in defence of the prevailing excise policies and one that remains in fashion. It is worth noting that using taxation to discourage alcohol consumption by raising its retail price to a desired minimum is currently an official policy in Canada and some US states, and plans to impose a minimum price per unit on alcohol have been hotly debated in the United Kingdom in recent years (Stockwell 2014). Modern debates emphasise the public health benefits of reducing alcohol consumption, but the basic mechanism of using taxation to limit consumption is nevertheless remarkably similar to the 'Maximum Revenue from Minimum Consumption' policy of colonial India. Although the excise officials of late nineteenth-century India probably would not have thought of their policies in terms of public health benefits—health concerns were limited in practice to imposing basic hygiene standards on distillers, preventing adulteration, and steering consumers towards weaker liquor—they certainly did emphasise the social and moral benefits of their policies and regulation. Their chief concern aside from revenue (one they shared with many of their critics) was that excessive consumption of strong liquor led to public nuisance and public disorder. In this volume, Dutta outlines the efforts made by the colonial state to protect the health of European sailors from the deleterious effects of 'noxious' local liquor in the major colonial port cities, notably in Calcutta. Outside these hubs of empire, poor Europeans were rare and there was very little official concern over the relationship between alcohol consumption and the health of the individual. Excise officials did occasionally voice concern that excessive drinking could injure and demoralise already impoverished families and could result in lost workdays. However, the excise bureaucracy also argued, quite reasonably, that complete prohibition would have been utterly unenforceable; the alternative would be what one official report described as a 'disastrous free-for-all' in the liquor trade. By this logic, every rupee collected by the excise administration was one rupee kept out of the hands of unscrupulous private traders or criminals and represented less alcohol consumed than would have been the case under either a free market or a repressive regime of prohibition that could never realistically hope to stamp out illicit distillation.

For this moral argument to remain valid, the excise bureaucracy needed to demonstrate that their policies brought order and stability to the liquor trade and enabled the state to better control access to potentially harmful intoxicants. Slow, steady increases in excise revenue were presented as clear signs of success—proof of the suppression of illicit liquor production or of a rising retail price and therefore a decline in per capita consumption. Statistical data showing rapid year-to-year fluctuation in excise revenue, however, was easy to interpret as a failure of the existing excise regulation to provide such stability—or worse, as evidence that 'Maximum Revenue from Minimum Consumption' was a mere rhetorical fig leaf providing thin cover for policies designed to push alcohol addiction for the sake of government revenue. Nevertheless, sudden increases or decreases in excise revenues were common given the number of factors that could influence excise revenues, positively or negatively. The excise reports cite dozens of such factors: favourable or unfavourable harvests; natural disasters; changes in underlying economic conditions; the opening or closing of nearby military cantonments or large public works projects; crime and disorder; temperance campaigns or political agitation both of which were often linked in the official mind after about 1895; the spread of European habits; auspicious or inauspicious years for marriage; and, as we shall now see, disease. Excise officials throughout British India repeatedly used all of these factors to justify unexpected and potentially damaging results. Disease, however, stands out among the others because of how frequently and, more importantly, how inconsistently it was used.

Alcohol and Disease: The Claims

The first unambiguous references to correlations between disease and excise revenues date to the late 1840s. The *Report on the Abkarry Revenue of the Lower Provinces for 1852–1853*, the first of Bengal's annual excise reports, lists three causes for declining excise revenues in the districts around Calcutta from 1849:

> first, a rise in the price of rice; secondly, the great demand for labourers of all kinds for the Arracan Coast; and thirdly, the prevalence of fever among the population during the last three years including the present. (Bengal Excise Report 1852–1853)

This assertion reflects similar claims from the 1850s and 1860s in two regards. First, it attributes the decline in revenue to 'fever', something a modern observer would likely define as a symptom and not a diagnosis of disease. Indeed, up to the 1880s most references to disease in the excise reports simply listed 'fever' or 'sickness'. Only thereafter were vague allusions to illnesses replaced by references to specific diseases like cholera and plague. Second, the passage in the 1852–1853 Bengal report quoted above is representative of how the vast majority of these vague mid-century references to disease were included as part of an assortment of factors, rather than standing alone. The 1876–1877 Bengal excise report, for example, attributed a substantial decrease in the excise revenue of the Chittagong District 'to the distress and sickness caused by the cyclone of 1876' (Bengal Excise Report 1876–1877).

Disease and environmental catastrophe often appeared side by side in the annual excise reports. The link between them was presented as self-evident, as was their combined ability to influence excise revenues. Indeed, it was considered noteworthy when a combination of, say, flooding and disease failed to produce the expected result. For example, the Governor General's official reply to the 1886–1887 Bengal excise report noted:

> Considering that the area supplied by [the central distillery in Hooghly District] was that which *suffered the most from inundation and sickness*, the Commissioner expressed his surprise, not at the fact of the decrease

> [in gallons of liquor sold], but that the decrease only amounted to 1,781 gallons. (Bengal Revenue Department, Resolution, October 24, 1887. Italics mine.)

Drought was also frequently linked to disease in excise reports. Thus, in a report for 1872–1873, the Commissioner of Excise for Bengal cited drought and fever as factors preventing the accurate assessment of a new system for licensing retail liquor shops:

> The Commissioner is of opinion that the result of the year under report scarcely affords sufficient materials for forming a decided opinion on the working of the new system for disposing licenses. There were several impediments such as the transfer of jurisdiction, *the scarcity of rain which led to a poor rice crop, and brought distress among the poorer classes, and also the sickness caused by the prevalence of epidemic fever*, but [the Commissioner of Excise] Mr. Buckland thinks that the new system will have the effect of diminishing the consumption of liquor and of increasing Government revenue. (Bengal Excise Report 1872–1873. Italics mine.)

Disease was also said to influence excise revenues, especially in reports from the 1860s and 1870s. This example from the 1869–1870 Bengal excise report is representative:

> The increased [liquor] revenue shown… in districts where the increase is noticeable, are ascribed generally to the prosperous condition of the people, and to improved administration and careful supervision of the central distilleries. The decrease in other districts is attributed generally to dearness of provisions from the scarcity, and to the comparative high price of mowah,[2] which affects the sale of spirit. Special causes are in some instances indicated; such as establishment of shops for sale of imported wines; diminution of employment on lines of railway; *prevalence of epidemic fever in Burdwan and Hooghly*, and to the greater consumption of other articles of excise. (Bengal Excise Report 1869–1870. Italics mine.)

Disease stands out from other factors cited in the excise reports for two reasons: it could have a dramatic, short-term, and unpredictable impact

[2] Fermented flowers of the *Mahua longifolia* tree were a common base ingredient for the production of so-called country liquor, prior to the widespread adoption of sugar or molasses by industrial distillers in the late nineteenth century. *Mahua* flowers remain a popular ingredient in homemade liquor wherever the tree grows in quantity.

over a very limited area; and, unlike most other factors, that impact could be positive *or* negative.

Other commonly cited external influences on excise revenue, such as the quality of the harvest, economic boom or bust, the beginning or end of major construction projects, famine or environmental calamity, and so on, had more or less predictable results. Favourable harvests or an influx of labourers were linked to an increase in excise revenues, whereas scarcity of food or natural calamities like floods had the opposite effect. Socio-cultural factors like auspicious years for marriages in communities prone to celebrating such events with liquor, or the picketing of liquor shops by local religious leaders, frequently took excise officials by surprise, but their impact on excise revenues was relatively straightforward to assess. By contrast, the impact of disease was more complicated: disease could result in lower than expected consumption if it ravaged the local economy, disrupted important festivals, or resulted in an exodus from the afflicted area, but the spread of beliefs in the preventative or curative properties of alcohol could also boost excise revenues under otherwise adverse conditions. Often, the impact of disease on excise revenues was utterly unpredictable or at least presented as such in official reports. One exceptional case reported in Bombay in 1891–1892 attributed a decline in excise revenues to the smuggling of illicit liquor from the Savantvadi state and Portuguese Goa; the liquor in question was 'manufactured from the juice of the cashew fruit' and, according to the report, was believed to have particular medicinal qualities the locally produced spirit lacked (Bombay Excise Report 1891–1892, 6).

The presence of endemic disease was also claimed to influence excise revenues, especially during the middle decades of the nineteenth century. Although they never clearly articulated the reasons, excise officials made frequent references to prevailing beliefs that people living in malarial environments derived health benefits from habitual alcohol consumption and that such beliefs explained the higher per capita rates of liquor consumption in many malarial areas. Belief in a relationship between alcohol consumption and health in malarial areas manifested itself in a number of interesting ways. One particularly intriguing reference comes from an 1867 report on the Chota Nagpur Division of Bengal:

> It is to be borne in mind that the wages of the labouring classes have not… risen in proportion to the increase in the prices of all the necessities of life, and the pittance received being barely sufficient to provide daily food for the

> workman and his family, he must forgo the luxury of drink. How far this may have affected the health of the poorer classes is a question well worth consideration; there can be no doubt that the mortality last year exceeded that of previous years, not only in parts of the Division that were considered as affected by the [Orissa] famine [of 1866], but also in Districts where there was little or no scarcity. *It has been found that a stimulant of some kind is beneficial to people who are much exposed to the influence of malaria, and it is not impossible that a malignant fever which, in addition to other epidemics last year*, assisted in keeping down the increase to the population, *might have proved less fatal, had the people been in a position to indulge as usual.* (Bengal Excise Report 1866–1867. Italics mine.)

This belief in the ability of potent alcohol to shield against malaria and fever seems to have been one that many mid-century excise officials shared. They entertained vague notions that specific intoxicants produced different results on people depending on their race, the climate in which they resided, or both. Such ideas lost favour during the final decades of the nineteenth century, as improved scientific understandings of the physiological effects of alcohol and recognisably modern identifications of disease filtered down to provincial excise officials. Moreover, while excise reports from 1880 increasingly referred to specific diseases, they continued to cite prevailing local beliefs in the medicinal value of alcohol—while taking care to note that they did not share such ideas.

Cholera and the Plague

Late nineteenth-century excise reports, in contrast, make many references to the impact of specific diseases, notably of cholera and the plague. Curiously, frequent references to cholera and its impact on excise revenues appear in the excise reports only from the late 1870s, despite a series of devastating cholera pandemics that ravaged British India from 1817 onwards. This must partially result from the nature of the excise archive itself. The Bengal Presidency was the first to begin preparing detailed, explanatory excise reports in 1852–1853, but most other provinces followed suit only from the 1870s.[3] Cholera was endemic in much of the

[3] It is important here to note that these annual excise reports were compiled for the benefit of other branches of government (and later as part of the public record) and therefore go further to explain and justify the actions of the excise department than ordinary internal correspondence and bookkeeping.

Bengal Presidency, including the Delta region of lower Bengal, and the confluence of the Ganges and Brahmaputra rivers—two of the most populous, and from an excise standpoint most lucrative, regions of the province (Dasgupta 2012). Mortality from cholera in these regions was high, but far less variable from year to year than in Bombay, the Central Provinces, Punjab, or southern Madras, where the disease was epidemic, rather than endemic. For this reason, cholera would have offered a much less effective justification for any dramatic fluctuations in excise revenue in Bengal than it did elsewhere, a fact that goes a long way towards explaining its absence from the Bengal excise reports.

Elsewhere, however, specific references to cholera and its impact on excise revenues abound from the earliest dates that detailed excise records are available. Despite this, the stated nature of cholera's impact on excise revenues varied. The coincidence of cholera and famine in Madras in 1877,[4] as in Bombay in 1877 and 1900, was linked to notable declines in excise revenues (Bombay Excise Report 1879–1880). Nevertheless, cholera's impact was not always negative. Sometimes, cholera outbreaks were said to have increased excise revenues. Consider this example from the 1887–1888 Madras excise report:

> The latter half of the year was marked by a sudden return to the rate of consumption of some years past in the towns and in the Talukas of Trichinopoly, Musiri and Kalitahai. This improvement (which was very marked) took place in October 1886, but what its true cause was in the taluka I am unable to conjecture. It has been attributed to the terrible cholera epidemic which prevailed during the last cold season. (Madras Excise Report 1887–1888)

This example is especially interesting as it represents one of the very few cases to be found where a claim regarding the effect of disease on excise revenues was officially disputed. The Commissioner of Excise, having done due diligence, dismissed the assertion, noting that he did 'not think that [cholera] had much to do with it, as the improvement was very noticeable two months before any case of cholera took place at all and still continues though the epidemic has nearly disappeared' (Madras Excise Report 1887–1888). Elsewhere, however, claims that cholera outbreaks resulted in increased excise revenues, as people who did not normally

[4] The 1866 cholera epidemic in Madras also coincided with a famine, but this predates the creation of a distinct excise department in that presidency.

consume alcohol turned to potent liquor for its supposed medicinal properties, were accepted without comment, as this example from Bombay illustrates:

> In the Deccan and Southern Maratha Country... The larger sales in all the districts mentioned were probably due to a more favourable season, the Collector of Satara assigning an additional reason, *viz., the prevalence of cholera in the district* during the early part of the year. (Bombay Excise Report 1894–1895. Italics mine.)

Plague occupied a similar position to cholera in official excise reports, in which officials used them to explain both increases and decreases in excise revenues. The examples cited at the beginning of this chapter need not be repeated, but as the arrival of plague to India in 1896 provoked a far more extreme reaction on the part of the colonial state than any of the earlier cholera pandemics, it is worth examining the different ways that plague and plague restrictions featured in the excise records.

Most references to plague were relatively straightforward. The initial appearance of the disease greatly reduced excise revenues. In the city of Bombay, as we have seen, the massive decrease in excise receipts recorded for 1896–1897 was said to have been the result of people fleeing the city. The revenues from Bombay recovered dramatically the following year, but that increase was offset by the spread of plague to rural areas. In Bharwar District, for example, the decline was attributed to 'poverty and distress consequent on the high price of food and spread of plague' (Bombay Excise Report 1897–1898). Likewise, poor excise revenues in what is now southern Gujarat were said to be the result of 'plague and scarcity'. The results were mixed in the districts close to Bombay. The substantial year-over-year decrease in Ratnagiri District was explained by the fact that fewer people had fled to that district from Bombay than in 1895–1896, but sales of liquor were up dramatically in nearby Thane despite similar circumstances.

Efforts to combat the plague were said to have a negative impact on excise revenues, often in complex, unexpected ways. For example, the 1898–1899 Madras excise report stated:

> [In Anantapur] the decrease is attributed to the *plague scare*, and the related paralysis of trade for fear of the various calamities predicted for the year... [In Chingleput,] *in consequence of the plague*, the attendance at most of the

> festivals has been much smaller than usual... [In South Arcot,] fairs and festivals were placed under considerable restrictions *on account of the plague* [and] in the early part of the year the season was adverse and affected the drinking classes especially. The district surrounds Pondicherry, where liquor is to be got cheap, and there is always a tendency which is accentuated in bad years for the people in the neighbouring British territory to patronize French shops... [In South Canara] the high prices of food grains and the poor demand for labour on coffee plantations have brought the drinking classes to poverty and debt; *plague restrictions interfered with* the usual large gatherings at the annual Hindu jatras; and the traffic between the district and Mysore, which entailed a large demand for liquor, has considerably fallen off. It is said also that there has been an increased resort to toddy and foreign liquor. (Madras Excise Report 1898–1899. Italics mine.)

The initial impact of the 1896–1897 plague pandemic on excise revenues was universally negative, as sickness, mortality, and the resulting panic severely disrupted all aspects of normal life. However, in subsequent years, excise officials returned to the familiar pattern of reporting mixed results. In Bombay, references to a prevailing belief that 'liquors act as a tonic or preventative against plague' began to appear in the excise reports from 1898, and increased excise revenues elsewhere in the Bombay Presidency and in Madras were attributed to the easing of plague restrictions even after revenues had surpassed their pre-pandemic levels (Bombay Excise Report 1887–1888).

The above examples illustrate the variety of ways that disease was said to impact the excise revenues of British India. Disease, it was claimed, did not have a single direct, predictable correlation with rising or falling liquor consumption or excise revenues. Whereas other factors had predictable results—for example, an economic boom always raised excise receipts, while the presence of a temperance campaign or a bad year for marriages always lowered them—disease was used to explain almost any positive *or* negative results, just as its appearance alongside other more predictable factors could be used to explain why they failed to have their supposedly predictable effect.

Alcohol and Disease: The Statistics

Thus far, this chapter has presented official claims about the relationship between excise revenues and disease at their face value. The truth of such claims was, insofar as the authors of the annual excise reports were concerned, more or less irrelevant. Trapped between the imperative to enhance revenue on one hand and avoid providing ammunition to increasingly vocal critics on the other, the authors of the annual excise reports only needed their claims to be plausible, not necessarily verifiable. The prevalence of references to disease and its impact on excise revenues in the annual provincial excise reports, as well as the remarkable lack of pushback against such claims in either official replies from higher branches of government or in the pamphlets and editorials produced by critics of the excise administration, suggests that the strategy was an effective one. However, did disease truly impact excise revenues in late nineteenth- and early twentieth-century British India? Although providing a conclusive answer to this question is not here the primary goal, as any attempt to do so conclusively would require a monumental effort to gather and assess data on a district by district basis, macro-level statistical data can still be used to test claims.[5]

Available data suggests that both the cholera pandemic of 1877 and the plague pandemic of 1896 slowed the otherwise steady growth of excise revenues throughout British India. The statistics also underscore the significance of the plague pandemic of 1896 and the subsequent imposition of associated restrictions on trade and movement throughout British India.

However, the large dip visible in Chart 7.1 coincides not with the 1896 plague pandemic, which resulted in only a slight overall decline in excise revenues, but rather with the devastating famine of 1899–1900. Cholera in 1877 and plague in 1896 both managed to temporarily arrest the otherwise steady growth that characterised excise revenues between the 1870s and 1910s, but the effects of the 1899–1900 famine were so great that excise revenues did not recover to pre-famine levels for nearly a decade. The famine loomed large in a way that it is impossible to understate, and for that reason, this chapter deliberately avoids including examples from after 1898–1899. The sixth cholera pandemic, beginning in 1899–1900, coincided with both the outbreak of famine and the large decline in excise revenues after 1899, but the two are

[5] The data used to produce the charts that appear in this chapter was compiled from two sources. Excise revenue data for Bengal, Bombay, and Madras comes from the provincial excise department annual reports, which can be found in the India Office Records Collection held at the British Library (IOR/V/24/1089-1093 for Madras; IOR/V/24/1098-1102 for Bombay; and IOR/V/24/1130-1135 for Bengal). Excise data for India as a whole and all disease mortality data are drawn from various editions of *The Statistical Abstract of British India*.

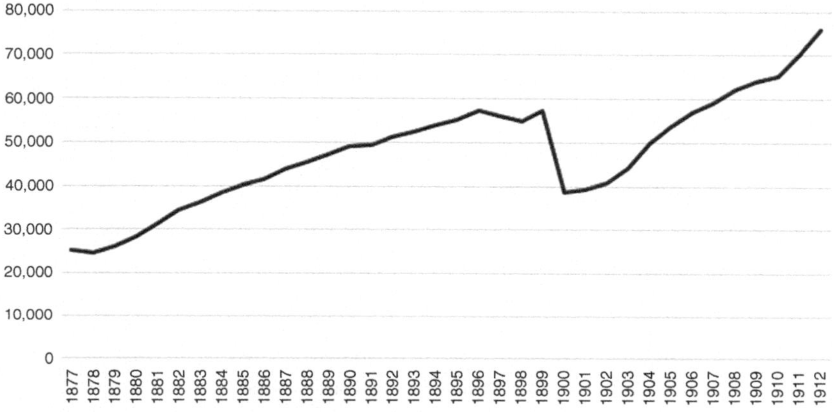

Chart 7.1 India excise revenue in thousands of rupees

impossible to disentangle. Furthermore, the immediate and devastating effects of the famine temporarily removed any need to offer detailed justification for declining excise revenues. Officials explained even the dramatic year-over-year increases in revenue experienced after 1902 as simply a gradual return to pre-famine levels.

A study of per capita rates of excise revenues on a provincial level in Bengal (Chart 7.2), Bombay (Chart 7.3), and Madras (Chart 7.4) also clarifies the relationship between revenues and disease.

On a provincial scale, the negative impact of the 1877 cholera outbreak in Bombay and Madras becomes clearer, but the plague pandemic of 1896–1897, a defining event if the annual excise reports are to be believed, becomes difficult to distinguish from other trends.

Charting excise revenue per capita of each province in terms of percentage annual change provides no better evidence for any major correlation between disease and excise revenues. The only noticeable event is again the famine of 1899–1900.

Comparing annual changes in per capita excise revenue to annual mortality figures from cholera, plague, and the nebulously defined 'fever' at provincial level also shows no strong relationship between excise revenue and disease (See Chart 7.5, below). Of the two worst years for cholera in Bengal between 1877 and 1919, one (1908) coincides with a large per capita increase in excise revenue, while the other (1900) does not.

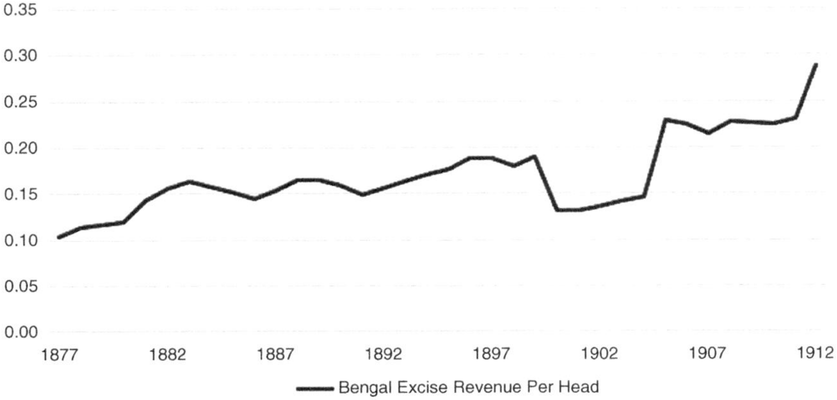

Chart 7.2 Annual excise revenue in rupees per head, Bengal Presidency

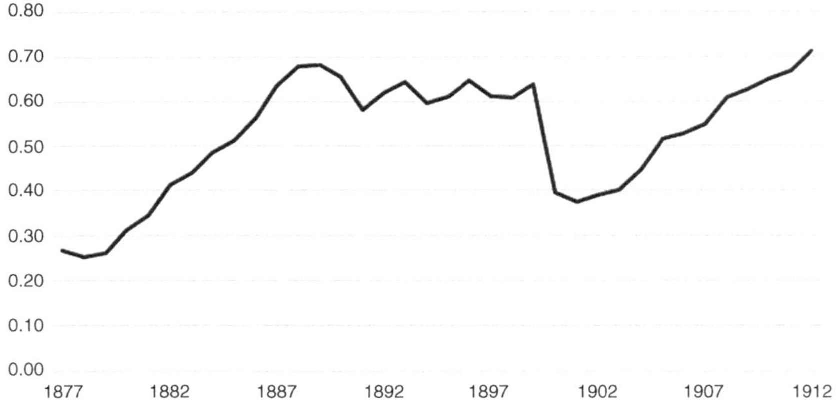

Chart 7.3 Annual excise revenue in rupees per head, Bombay Presidency

Unfortunately, in Bengal, the initial outbreak of bubonic plague in 1897–1898 is not properly captured in the statistics, but it does appear to coincide with a rare decline in per capita excise revenues. The worst subsequent years for plague were 1901, 1906, and 1911. Of these, 1901 and 1906 saw almost no per capita excise revenue growth, whereas 1911 marked the beginning of several years of extremely rapid growth that would only be arrested during the First World War. 'Fever' predictably

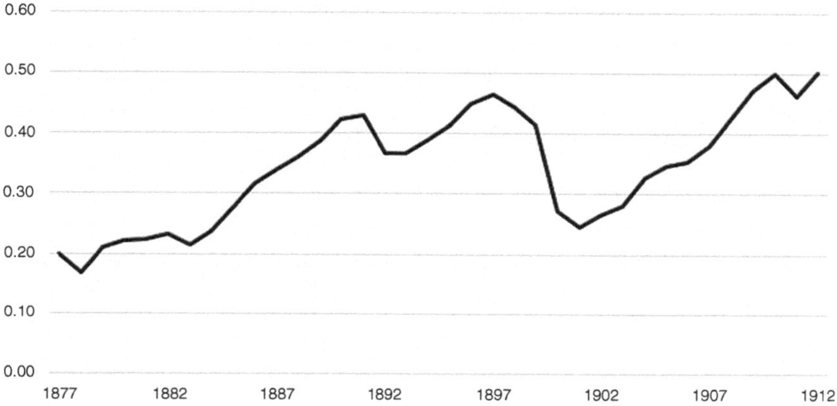

Chart 7.4 Annual excise revenue in rupees per head, Madras Presidency

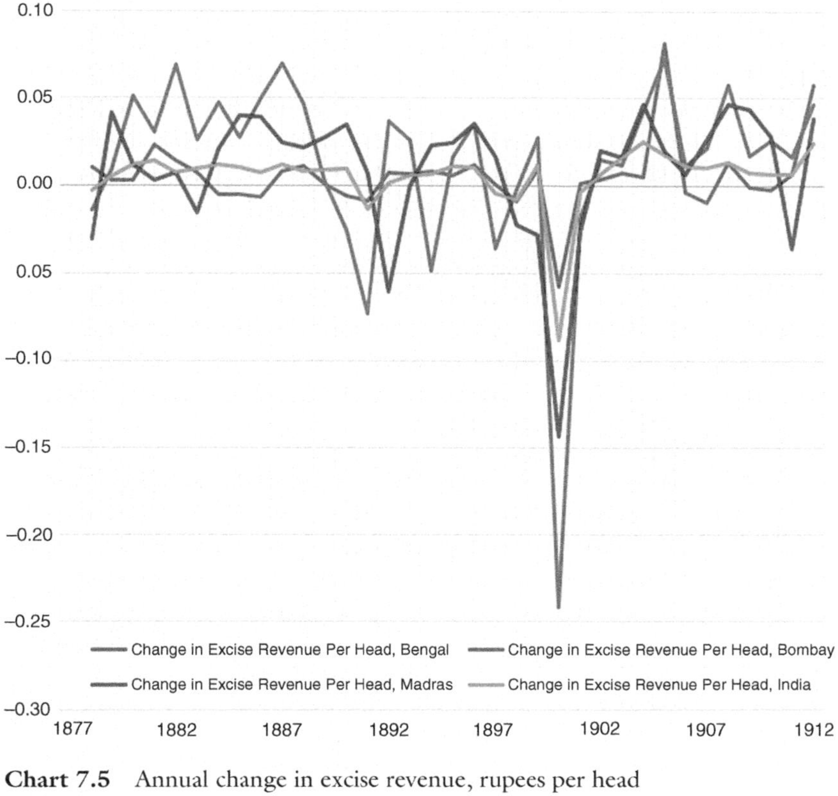

Chart 7.5 Annual change in excise revenue, rupees per head

shows little or no correlation with excise revenues. In Bombay, the initial appearance of the plague at the end of the nineteenth century did coincide with a period of lower than normal excise revenue growth, and although a reprieve from plague coincides with a year of exceptional growth in 1905, the relationship between plague and excise revenue remains minor, at best. The major cholera outbreak of 1900 coincides with a decline in excise revenue in which, however, as noted above, famine certainly played a much larger role. Finally, in Madras, mortality from cholera does not appear to have had any strong impact on excise revenue growth. The worst period for plague in the presidency, 1901–1905, did coincide with a period of excise revenue growth, but the return of plague in 1911 marked a low point for excise revenue growth in the Madras, in contrast to Bengal.

Conclusion

While the above analysis cannot and does not claim to be exhaustive, it is clear that, at least at the macro level disease had no consistently serious impact on excise revenues in late nineteenth- and early twentieth-century India. Why, then, was disease so often invoked by agents of the excise department to explain and justify unexpected results? The answer, it seems, was that such explanations *worked*. Disease offered a multitude of plausible explanations for increases or decreases in excise revenues. Outbreaks of epidemic disease were especially useful in justifying dramatic and unexpected increases or decreases in excise revenues because they represented discrete events that were both simultaneously unpredictable and unpreventable, and of sufficiently short duration to demand no modification to carefully tuned systems for regulating alcohol production and collecting excise revenues. Readers of the official excise reports—government agents and external critics alike—do not seem to have questioned these alleged links between disease and excise revenue. It is not the purpose here to investigate whether such claims represented true accounts of minor, localised phenomenon, wishful thinking on the part of the officials involved, unintentional reporting of false information, or deliberate attempts to deceive. Its aim, in the context of this volume, is to suggest one of the ways that the disease environment of the Indian Ocean World, and especially the major disease outbreaks of the late nineteenth century, could have had a subtle impact on forms and processes of colonial governance. The actual links between disease and the smooth collection of excise revenue in British India may or may not have been real—but to the excise officials reporting them, that hardly mattered.

References

Primary Sources

India Office Records Collection

IOR/V/24/1090-1093. Madras. Excise Department: Report on the Administration of the Excise Revenue in the Presidency of Fort St. George. (1881–1928)

IOR/V/24/1098-1102. Bombay. Excise Department: Report on the Administration of the Excise Department in the Bombay Presidency (1879–1922)

IOR/V/24/1130-1135. Bengal. Excise Department: Report on the Administration of the Excise Department in the Presidency of Bengal (1852–1922)

IOR/L/PARL/2/287-292E. Statistical Abstract relating to British India. Nos. 13-55 (1877–1920)

Secondary Sources

Dasgupta, Rajib. 2012. *Urbanizing Cholera: The Social Determinants of Its Re-Emergence*. New Delhi: Orient Blackswan.

Hynd, Peter. 2020. *Excise and Alcohol in British India*. Thesis, McGill University.

Stockwell, Tim. 2014. Minimum Alcohol Pricing: Canada's Accidental Public Health Strategy. *The Conversation*, April 4. Accessed 15 Feb 2018, https://theconversation.com/minimum-alcohol-pricing-canadas-accidental-public-health-strategy-25185.

CHAPTER 8

European Sailors, Alcohol, and Cholera in Nineteenth-Century India

Manikarnika Dutta

Introduction

In early modern Europe, British sailors had the reputation of being rootless, often violent, promiscuous, and dipsomaniacs—an idiom for social evil (Conley 2009, 2). However, in the late eighteenth century, this image started to change as naval authorities, Christian missionaries, authors, and playwrights sought to project sailors as valiant founders of Britain's maritime empire. Subsequently, in Victorian and Edwardian Britain, sailors were increasingly portrayed as defenders of the nation and devoted family persons. American sailors were likewise presented sympathetically in accounts such as the 1852 annual report of the American Seamen's Friend Society, which stated that more than 70,000 sailors had pledged themselves

Research for this chapter was done as part of a doctoral fellowship from the Wellcome Trust-funded project 'From Sail to Steam: Health, Medicine and the Victorian Navy', held at the Wellcome Unit for the History of Medicine, University of Oxford.

M. Dutta (✉)
University of Oxford, Oxford, UK

G. Campbell, E.-M. Knoll (eds.), *Disease Dispersion and Impact in the Indian Ocean World*, Palgrave Series in Indian Ocean World Studies, https://doi.org/10.1007/978-3-030-36264-5_8

to the temperance movement and nearly 50 Sailors' Homes had been established in port cities (Haughton 1855, 111). At the same time, the merchant and war navies of several countries supplied sailors with premium quality coffee and tea in an attempt to discourage them from drinking liquor. Naval authorities around the world floated the idea that abstinence improved the chance of surviving the ordeals of maritime voyages (Gustafson 1884, 97). However, in stark contrast to this generous characterization of the sailor in Europe, the British Indian government expressed great anxiety about the worsening situation of European sailors in Indian port cities. Its efforts to regulate the health and behaviour of sailors drew a fault line between imperial and colonial contexts of governance.

Alcohol was widely considered central to seamanship, to the extent of being a determinant of the collective behaviour of sailors. Drinking together was a means to befriend strangers, allay fears of perilous journeys, and generate a sense of bonhomie among crew members. The captain exercised authority over his crew through regulating their access to alcohol (Gray 2016). Consumption of liquor by discharged sailors at port cities at the end of a ship's journey was more problematic. In early nineteenth-century India, members of the British ruling elite were particularly embarrassed by their failure to adequately discipline drunken sailors (Tambe 2009). They were also concerned that local liquor, often found to be adulterated and of low quality, was pernicious enough to cause Europeans to become debased. Harald Fischer-Tiné has argued that colonial narratives blamed crime among sailors on the consumption of drugged liquor rather than pure European alcohol. He observed that the behaviour of these drunken 'white subalterns' embarrassed the colonial government, which considered sailors to be a potential threat to its policy of racial supremacy. The state was concerned that if the reckless behaviour of European sailors brought them closer to the 'uncivilized natives', its lofty ideals of the civilizational purity of race and class might implode (Fischer-Tiné 2012). This anxiety was visible in colonial policies regarding the health and welfare of seamen.

The works of Fischer-Tiné and David Arnold (1979) have explored the disreputable section of European settlers in India that the colonial state was unwilling to acknowledge as its own. Douglas Peers and Erica Wald have analysed the colonial state's efforts to maintain the health of European troops by regulating their intake of liquor. Peers (1998) contends that British military administrators were concerned at the high morbidly rate among European troops of both diseases such as cholera, malaria, and plague—almost twice that of Indian sepoys—and of venereal diseases, which they associated with alcohol consumption. The authorities were initially reluctant to prohibit soldiers from venturing out of the barrack for sexual gratification, lest it led to an increase in alcoholism and homosexuality in the barracks. Later, Act XVIII of 1853 enabled commanding officers to prosecute unregistered liquor dealers who were previously untouched by military law (Peers 1999). Wald's (2012) analysis of military records reveals that, unable to control the alcohol consumption of soldiers, garrison authorities tried to restrict alcohol suppliers. They criticized the wives of soldiers who practised prostitution and procured alcohol illegally, and the arrack vendors in the vicinity of barracks who presumably sold adulterated liquor. The works of both Peers and Wald, in emphasizing the colonial state's need to preserve order and health among its soldiers, focused on the administrative strategies of limiting alcohol consumption within the military establishment.

Building on previous scholarship, this chapter explores the impact that adulteration of liquor and the 'crimping system', in a colonial setting, had on the health of white sailors who British authorities in India believed to be vulnerable to local influences such as cholera and cheap liquor. Additionally, liquor was envisaged as a cause of cholera at a time when aetiologists were uncertain about the exact nature of bacterial contamination leading to the disease. In the nineteenth century, port cities were known as disease entrepôts where disease pathogens could be spread both towards the port's hinterland by road and rail links and outwards through oceanic voyages. As Arnold (1991, 10) has argued, as an 'incubator of infectious disease' like cholera, Calcutta was at the focus of health hazard monitoring. Soldiers, sailors, pilgrims, and migrant labourers passing through the city were potential agents of disease dispersion. Although whether drinking or drunkenness constituted a disease was largely a matter of opinion rather than medical fact, a number of surgeons suspected a connection between liquor and cholera. The latter was still considered a disease that originated in a specific locality, that is, the lowlands of Bengal. The nature of cholera was so little known until at least the mid-nineteenth century that physicians could not reconcile various theories of cholera occurrence and naval surgeons were often unable to distinguish it from severe diarrhoea and dysentery (Preston 1895). Thus, consumption of liquor was a potential threat to public health that could escalate into a massively fatal but little understood epidemic.

Thus, this chapter argues that in British India sanitary regulation was inseparably linked to an anti-vice attitude in administrative policies. It begins with an analysis of the Government of Bengal's attempts to regulate liquor consumption among European sailors. Thereafter it examines the connected history of the fight of British colonial officials against adulteration of liquor, notably in Calcutta. Finally, it traces the adverse social effects of the medical problems of adulteration on the behaviour of sailors in port cities. The chapter aims to investigate the extent to which the colonial state's measures to protect the health of sailors were informed by imperial encounters in the fields of medical intervention, race relations, environmentalism, and legal order. Specifically, this study of adulteration and quality control of drinks takes into account David Arnold's (1993) emphasis on the linkages between colonial and British medical developments. Shula Marks (1997) has pertinently pointed out the necessity of considering what precisely constituted 'colonial' in the overlap between the colony and the metropole. She specifically points to the desire of British colonizers to uphold certain ideologies of governance and the differences they considered to separate the colonizer from the colonized. This forms the context here for an examination of the agency of colonial power in controlling the discourses of medicine, in which emphasis is placed upon the extent to which medicine in the colonial situation differed from the metropolitan context.

Prevention of Mortality and Regulation of Profligacy Among European Sailors

Colesworthy Grant (1813–1880), a British resident of Calcutta, wrote in 1850 that white sailors brought immense disrepute to the city. Their favourite haunt was Flag Street in Lalbazar, which was full of taverns run by Indians and Europeans from Italy, Spain, and Portugal. They would 'disappear for days in the nests and fastness of riot and profligacy which are numerous in that vicinity' (Barrett 2004, 384). In order to retrieve their sailors in time for the outward journey, captains of ships often had to visit gambling and drinking dens, sometimes accompanied by police constables. As part of a global campaign, the temperance movement in Britain inspired voluntary organizations in India, especially Christian associations, to discourage sailors from 'temptations'. When stories of drunk European and American sailors in Indian port cities circulated internationally, the

British colonial state became active in disciplining them. The police were always on the lookout for 'houses of ill-repute' that sponged off visiting sailors. Since they did not expect abstinence from the sailors, they monitored the sale and consumption of liquor, particularly adulterated liquor, in an attempt to restrain and protect their health.

In November 1832, the magistrate fined Charles Neville, Richard Barrett, and Thomas Owen for running unlicensed public houses. On the same day, a police sergeant named Crawford broke into an illegal tavern maintained by a person named Lazarus. He had been spying on a particular den for months, but somehow Lazarus had been evading his search. Finally, when Lazarus was apprehended and brought before a court, he denied the accusation, saying that he was ill and had little idea when and how the sailors occupied his living room as if it were a tavern. He added that the sailors ordered food and drink without him noticing, and had no inkling about the bottles of alcohol that the sergeant had seen on the table. He was found guilty and fined 15 rupees (*John Bull*, 4 November 1832). Around the same time, about a dozen drunken sailors assaulted two Indian police constables opposite the Writers' Building. They wanted to take the rattan stick wielded by one of the constables and started beating the constables when they were refused. A third policeperson who came to his associates' rescue was beaten as well and later taken to the Native Hospital. The main perpetrator stumbled against a wall and fell in a drain as the sailors gambolled towards the river. He was detained until a European sergeant apprehended him. The other sailors were summoned to appear in court (*John Bull*, 25 November 1832). According to an observer (Sykes 1992, 47–48) writing in the 1850s:

> There is a great outcry in Calcutta, and for once a reasonable one, against the grog shops, and the danger of them to the British soldiers and sailors. The spirit is bad, and very cheap indeed, and they have indulged terribly in it. Some got drunk that their medals were robbed from them, and few have died of drink... A good many days ago, the Lieutenant-Governor [of Bengal] was told to enforce the act withdrawing licenses from those shops where people came out drunk, but now a better thing is being done by establishing a Government canteen on the Maidan in tents, where good spirits and tea and coffee and beer can be had, and skittles and games, and newspapers and books for amusement.

Dr. Hugh Macpherson (1861), Inspector General of Army Hospitals, noted that between 1856 and 1860, 716 European Protestants died in Calcutta from cholera. Of these, some 76 per cent were part of the city's floating population, mainly sailors. Macpherson remarked that ship captains and crews had identified certain anchor sites along the Hooghly River as more perilous than others. Colvin's Ghat, close to the mouth of a long sewer, was particularly notorious, as were Thompson's Ghat, Cooly Bazar, Fort Point, and Armenian Ghat. He concluded that, though none of these sites were perennial breeding grounds of diseases, disembarking sailors were in danger of catching cholera. A report in the *Saturday Review of Politics, Literature, Science and Art* (29 April 1865, 507) noted that 11 cholera victims were admitted to hospital from a house in Bow Bazar within a few weeks. It criticized the Calcutta Municipal Corporation as a body of 'wrong-headed people who muddle each other and do no earthly good for public'. The lack of progress in combating cholera is evident from a report in *The Lancet* (5 November 1887, 931–932) which blamed the high mortality among European sailors, around 11.1 per 1000 persons, on 'breathing [the city's] foul air, and partaking of drinks diluted not always with hydrant water'. It noted that the Jack Tar was expected to adapt to this unparalleled atrocious environment, portraying them as helpless victims of circumstances unique to the colony.

The Government of India was arguably more concerned with the health of European soldiers than sailors. The *First Annual Report of the Sanitary Commission for Bengal, 1864–1865* contained 55 pages on the need for improving the diet and barrack accommodation of troops, and the necessity of new hospitals for them, compared to only three pages on sailors. A *Friend of India* correspondent (1863) wrote that sailors were neglected in comparison with soldiers, whose achievements were celebrated vigorously, especially after 1857, in government blue books, newspapers, and pamphlets. Death and disease in the army were given more importance. Questioning the colonial government's policy, the newspaper correspondent asked:

> What does India do for the sailors who carry to and fro the wealth which enriches her? Nothing that can be appreciated. Calcutta, Bombay and Madras are all bad alike and all complain equally… choleraic drains, a life-destroying sun, drugged brandy, brothels exceeding in beastliness the pictures of juvenal, robbery under the name of discount and charge on bills and notes.

It was the combined effort of actors in Britain and India that finally made the colonial state ponder the need for greater medical attention for sailors. An address by Dr. Norman Chevers (1818–1886), Principal of Calcutta Medical College, to seamen at the Floating Library, Calcutta, on 5 January 1864, provided the initial impetus for the state to act. Lamenting the poor physical state of sailors who should be healthy and convivial, Chevers (1864) wrote, 'The British Seamen ought to be—and, when placed under favourable circumstances, is,—one of the healthiest of mankind'. While the state was aware of the high mortality rate among European sailors, it was only after the publication of Chevers' essay that it began taking measures to improve the living conditions of seafarers in Calcutta. State officials started maintaining registers for sailors and enacted new sanitary regulations on the basis of accumulated data. As part of its welfare programme, the Sanitary Commission recommended providing sailors with comfortable accommodation and amusement (Cave-Browne 1865, 466). Sanitation and hygiene in the old Sailors' Home, constructed in 1838 in Bow Bazar, had by then deteriorated.

A second impetus for state action was the cyclone of 5 October 1864 that destroyed many ships at the port of Calcutta, leaving 547 European sailors without occupation. This meant that, as 458 discharged sailors were already on shore, the port authorities had 1005 sailors to rehabilitate. Many of them landed, destitute and sick, and ended up in prisons and hospitals. By 23 January 1865, the authorities had provided for 563 sailors, sending home 163 at the expense of the Board of Trade and the Relief Fund and 187 on nominal wages, recruiting 30 for the Royal Navy in Bombay, and employing 183 others on full wages, chiefly in maritime activities (Annual Report on the Administration of the Bengal Presidency, 1864–1865, 102). Further, the Lt. Governor of Bengal asked the Sanitary Commission to investigate the state of sailors. In May 1866, Major G.B. Malleson (1866, 9) responded with a report titled 'The State of Sailors in Calcutta', raising concerns about their living conditions, health, and conduct. He quoted the comment by the Superintendent of the Reserve Force of Police that Calcutta port was overpopulated with sailors. Another report from Captain Alexander Caw, Shipping Master, showed that between 1 May 1864 and 30 April 1865, 629 ships with a total of approximately 17,298 sailors visited the port of Calcutta. Of this number, 3655 were discharged, 129 deserted, 214 were sent to prison and 232 to hospital, 231 died, and the rest were left without employment (Malleson 1866, 4).

Malleson (1866, 5) indicated that Caw wrote a letter to the Board of Trade on 30 June 1865 expressing concern about the Calcutta port turning into 'a sort of a depot' for seamen from other ports of British India such as Bombay and Rangoon; from Shanghai, Sydney, Melbourne, and Port Louis (Mauritius); and from English towns such as Shields and Tyne. This influx strained the ability of Bombay port authorities to employ and accommodate sailors. Consequently, they stipulated, in the 206th section of the Merchant Shipping Act, that captains should contact the Shipping Master before discharging any inbound sailor and that they could be prosecuted for mistreatment of employees should they ignore this ruling. Caw further stressed the need for the Board of Directors to prevent ships arriving from colonial ports such as Melbourne and Sydney, or ports in England, from discharging sailors in Calcutta unless those sailors possessed a contract guaranteeing their return passage. In support of his argument, he pointed out to the Master Attendant, J.G. Reddie, that on 12 July 1865 the number of jobless seamen in Bombay was 692, whereas the maximum the port could sustain was 500.

The surplus population of seamen further compounded their health and legal problems. Disease, suffering, and mortality among sailors were principally associated with poor eating habits, stale and contaminated air in tiny ship cabins, exposure to various unhealthy climatic conditions (Harrison 1999), and above all exposure to drunkenness and venereal diseases contracted from local prostitutes (Chevers 1864, 37). Chevers (46) stated that the Sailors' Home was 'surrounded with drinking shops of vilest description' and situated in the 'centre of about the worst atmosphere discoverable in this unsavoury city'. He probably meant a combination of insanitary living conditions and availability of 'vices' as the worst atmosphere. Chevers suggested the construction of a larger building in a 'healthier' and 'reputable' part of the city. Montague Massey (1918, 89), a civil servant, wrote that the Sailors' Home in the 1860s was a 'crying scandal', because it was situated in an area abounding with 'native grogshops in which [shopkeepers] sold to the sailors most villainous, poisonous decoctions under various designations' and 'boarding houses run by a thieving set of low-caste American crimps'. Moreover, Lalbazar, the hub of watering holes and brothels, did not have a working sewage system. The drains were mostly open and full of black putrid slime that had accumulated over the years. British travellers usually disdained the marketplaces for their 'disgusting' appearance (Chakrabarty 1991).

Newspapers such as the *Saturday Review of Politics, Literature, Science and Art* and *The Friend of India* criticized the irregularity of sanitary supervision and failure to enforce sale of hygienic food in the port area. However, the threat of fatal diseases failed to deter sailors from spending time there, mainly for want of better options. Malleson (1866) indicated that conditions in the Flag Street neighbourhood would encourage the rapid circulation of epidemics and demanded constant care and vigilance by both the police and the municipality. In a letter dated 25 February 1864 (British Library IOR/P/433/52:1866), S.H. Robinson, Secretary of the Sailors' Home, requested Lt. Colonel H.C. James, Private Secretary to the Lieutenant-Governor of Bengal, to provide a new establishment in a better locality. He also expressed the need for a recreation ground for seamen, an area enclosed with a bamboo fence, resembling a cricket ground. In response, the government enclosed a part of the *maidan* (a vast field between the fort and the esplanade) for sailors to play cricket.

The existing Sailors' Home was later sold, and the proceeds were used to build a new house at 13 Strand Road under the 'special care' of Lord Henry Lawrence (1811–1879) (Firminger 1906, 163). It was situated in a better locality but accommodated fewer than the 200 men housed in the former building. Henceforth, avoidance of disease and other social evils often determined urban restructuring. Chevers (1864, 51) recommended the implementation of better drainage along Flag Street to improve sanitation as a necessary measure to protect sailors from diseases such as cholera and dysentery. He further advised that, at the start of each cholera season (which usually lasted from July to early October), the captain of every vessel should be given a set of regulations to prevent the occurrence of the disease and to cure those crew members it affected. The authorities also thought about accommodating homeless sailors in other boarding houses, which were to be carefully inspected for cleanliness in order to ensure proper living standards. Landlords were warned that they might have their licences revoked unless they resolved problems reported by residents. However, the infamous liquor addiction of sailors proved to be a greater concern for the colonial state, particularly as the quality of liquor was below standard and sometimes proved fatal.

Drunkenness and the Abuse of Adulterated Liquor

As the number of deaths from cholera among European sailors increased, the quality of drinking water came under scrutiny (Macpherson 1861). The authorities believed that, in the absence of clean water, the intake of unfiltered river water was responsible for the high cholera mortality rate (Chevers 1864, 41). However, European doctors considered that there existed aggravating factors other than insanitary living conditions or poor quality water—namely, environmental factors like the 1864 cyclone (Gastrell and Blandford 1865, 126). They also commonly believed that consumption of adulterated alcohol might have been responsible for the prevalence of cholera in certain localities, particularly around the port. The Medical College Hospital, situated near Flag Street in the centre of the city, admitted more than twice as many sailors as the Presidency General Hospital, located at the city's southern fringe. Chevers (1864) estimated that as many as 10 per cent of the sailors entering the Calcutta port every year died of cholera.

It could be argued that the colonial state's concern about adulteration was influenced by the metropolitan British attitude. The chemist Fredrick Accum (1769–1838) discussed in 1820 how some dealers adulterated food and drink with harmful substances. 'There is death in the pot', he wrote, as a preamble to the detailed exploration into how the poisonous extract of *cocculus indicus* (popularly known as the black extract) was mixed with malt liquors as a cheap way to increase the level of intoxication (Accum 1820). Sometimes a substance called multum that was comprised of gentian root, liquorice juice, and black extract was used (Accum 1820, 6). More dangerous was the practice of adulterating wine with lead to stop the process of acescence and maintain the transparency of white wine when it became turbid. Even a small amount of lead acted as slow poison, prompting Accum to castigate those responsible as murderers (Accum 1820, 102–103). The book aroused considerable public attention, and in 1851 Thomas Wakley (1795–1862), surgeon and editor of the medical weekly, *The Lancet*, and his colleagues started a campaign about the dangers of adulteration. They observed under a microscope foodstuff bought from different markets. This was followed by the establishment of the Analytical Sanitary Commission under the supervision of British physicians Arthur Hill Hassall (1817–1894) and Henry Letheby (1816–1876). Hassall examined about 25,000 food and drink samples between January 1851 and December 1854. The ensuing report, published in *The Lancet*, emphasized many instances of death, poisoning, paralysis, or any illness caused by intake of adulterated drinks and attracted much attention (Morton 2005).

A Select Committee established in 1855 to enquire into the adulteration of food and drink advised that adulteration must be stopped in order to protect public health, honest merchants, consumers, and especially public morality—so as to prevent depreciation of 'the high commercial character of this country... both at home and in the eyes of foreign countries...' (Hassall 1861, 37–39). However, many producers and retailers claimed that adulteration was harmless and served consumers by keeping prices low (Otter 2006, 520). Nevertheless, in 1872 an amendment of the Adulteration of Food and Drink Act (1860) was passed, incorporating Hassall's proposal to appoint a public analyst, which resulted in the establishment two years later of the Society of Public Analysts. The 1875 Sale of Food and Drugs Act stipulated that manufacturers print a guarantee of purity on wrappers and packets alongside certificates obtained from public analysts (Morton 2005, 170–171). This measure was not adopted in British India until the twentieth century when the technology of ascertaining dilution levels first became available across the territory. As an essential aspect of sailor welfare, a pattern emerged in major administrative centres such as Calcutta, Bombay, and Madras of investigating allegations of adulteration and implementing crackdowns on public houses. In addition to concern for public health, checking the loss of revenue was a compelling reason for the government to control locally produced liquor. The East India Company monopolized the liquor trade in 1773 and subsequently generated huge revenue from the steep excise tax on alcohol. However, this chapter concentrates on the medical aspect of the state's intervention in liquor trade.

The threat of drunken sailors was not exclusive to Calcutta. In a letter to *The Mariners' Church Gospel Temperance Soldiers' and Sailors' Magazine* (December 1845, 542), G. Drago, Aqueduct Sergeant in Poona, reported preparing a petition for the government to order hotels and taverns to stop their entertainment programmes on Sundays. Sailors frequented these establishments and surrendered to the 'most vicious kinds to intemperance', creating 'disgusting and demoralizing scenes' that disgraced Britain. Michael Kirwan Joyce (1854, 1) of the Bombay Police noted that certain areas of the city, such as Dobee Tank and Duncan Road, had a high number of liquor shops, although these were dispersed rather than concentrated in one location, making it difficult for the police to raid. Many of these shops employed discharged European soldiers as unwaged crimps or 'catchpoles'. These people earned their livelihood from the 'plunder of the unfortunate wretches', such as sailors. The liquor shops also employed abandoned women, nominally as domestic servants, but in reality as prostitutes who lured sailors into their establishments, called 'Tereerams', and there drugged them with adulterated liquor, and looted their belongings.

Joyce (1854, 3) argued that adulterated liquor was responsible for numerous deaths. Many licensed shops sold concoctions exclusively for sailors. Named 'Sailor Jack' or 'Tom's Brandy', such drinks were prepared from strong arrack mixed with the 'poisonous juice' of *datura*, extracted from tobacco and chillies and opium. The beer, priced at 50 paisa per bottle, was a mixture of beer, water, and a concoction of vinegar, soap nuts, sugar, and soda. The so-called wine, which cost a rupee a bottle, was a combination of vinegar, sugar, Parsee Brandy, and a decoction of logwood. The abundance of liquor shops where such products were sold compounded the problem. Of the 422 liquor shops, 3 described themselves as hotels and 13 as taverns. The rest comprised 172 retail shops and 234 toddy shops (Joyce, 4).

The debates and correspondence about adulterated liquor engendered a discourse about the corrupting influence of the Orient on western people. Some colonial officials reported a conspiracy by the 'deceitful', 'cruel', and 'dangerous' natives to induce innocent white sailors to get drunk so that they could steal their possession. This narrative recast white sailors as victims of colonized subjects whose crimes, including theft and murder, were often attributed less to a lack of morality than to the intake of pernicious Indian narcotics mixed in drinks. Some reports excused sailors as semi-educated men whose only escape from hard physical labour was to indulge in liquor.

Official concern over the quality of liquor sold at local grog shops grew with the increase in reports of 'treacherous' Indians tricking 'innocent' European sailors into drinking 'poisonous' liquor. Colonial officials increasingly felt that they had to contend not only with the traditional ill-discipline and disreputable behaviour of mariners but also their consumption of adulterated liquor. This led the government to investigate the quality of liquor sold in local markets. In general, it was made 'fiery hot' with red pepper and other 'tongue-rasping' and 'bowel-scorching abominations', and in some shops was found to have traces of several strong narcotic drugs such as *datura*, *cocculus indicus*, and *gunjah* (Chevers 1864, 41). Investigations into the sale of adulterated liquor to poor

Europeans, especially sailors and soldiers, in Lalbazar, Bow Bazaar, Rada Bazaar, and College Street, revealed that many shopkeepers used 'native' liquors such as *Mudut* and *Doasta*, described by Chevers as 'bazaar sharab', and falsely sold them in English bottles with labels such as 'Old Tom' and 'Exshaw's Brandy' (Chevers 1864, 65). Better quality liquors such as 'Exshaw's first class Brandy' were sold to Europeans at a higher than market rate. Chemical analysis of liquor in Calcutta in 1857–1858 found that it was diluted rather than drugged, and it was less its quality than the quantity of consumption that resulted in drunkenness and associated misdeeds.

According to Chevers (1864, 37), unadulterated liquor was so difficult to obtain in Calcutta that a 'sober man' (ostensibly a British sailor) could hardly ever find proper beer. In many cases, after drinking liquor purchased from local grog shops, sailors contracted cholera from which they suffered a higher mortality rate than others affected by the disease due to their overindulgence in poisonous liquor (Chevers 1886). Chevers argued that the mortality rate could be reduced by a third if alcohol consumption was properly regulated. He speculated that equal duty on rum and *doasta* would increase the sale of rum and advocated that all *doasta* in Calcutta should be distilled under strict surveillance. However, despite the bad publicity given to it, low-quality 'poisonous' liquor continued to be sold unabated. In his report, Malleson (1866, 2) mentioned a conversation with Dr. C. Fabre-Tonnerre, the municipal health officer in Calcutta, who told him that Magistrate Macleod Wylie had reduced the sale of 'noxious liquor' in the late 1850s by raiding public houses and withdrawing licences from those selling tainted liquor. Disease, destitution, and crime were thus intermingled in medical reports.

A number of prominent people, organizations, and publications supported the campaign against adulterated liquor. The *Indian Year-Book for 1862* (Murdoch 1863, 117) applauded the newspaper *The Friend of India* for doing 'good service by directing attention to the alarming increase in the consumption of spirits'. It quoted the following from the newspaper:

> As if it were not tough that drunkenness should be the national crime of the English at home, and should only too unmistakeably characterise her sailors and lower classes abroad, it would seem as though the Government of India were determined to make their heathen subjects and their own soldiers as bad as the people of the mother country [...]

Since it proved difficult to dissuade European sailors from drinking, alternative steps were taken to ensure 'none but pure alcohol would be sold'. Additionally, the civic authorities tried to reduce the number of shops that sold the cheapest liquor on the grounds that 'sober men' should be provided the opportunity to buy 'well-made coffee', 'good soda water', 'ginger beer', and 'lemonade' at proper rates (Chevers 1864, 51). In a letter dated 13 June 1864, Chevers recommended appointing a competent official to inspect every ship arriving at the port and install special taps to sell good quality spirits, wine, beer, and other drinks to sailors.

Borradaile, Schiller, and Co., a major partner in the Port Canning Land Investment, Reclamation, and Dock Company in Calcutta, suggested sellers of 'that most intoxicating drink, the Indian Rum and Doasta' be banned and pay heavy penalties. They supported the reduction of duty on European drinks to make their price competitive in the local market. The municipality adopted Tonnerre's suggestion of registering all seamen on arrival in the Calcutta port and obtaining from the captain of each vessel a list of the number and cause of casualties on board (Malleson 1866, 3). The government closely monitored inquiries made by Chevers and other officers about 'unwholesome food and drink'. Chevers' report in particular drew much attention and led to appropriate measures for the first time. Following its publication in 1864, a special committee was appointed to assist health officers to systematically inspect food and drink sold in the public markets and confiscate 'unwholesome articles' (Chevers 1864, 63). However, such measures were not uniformly successful. Authorities in Bombay were more effective in regulating the production and sale of liquor, as Peter Hynd's chapter in this volume shows. He argues that in the late nineteenth century, the Bombay excise officials were not driven by the perceived public health benefits of regulation of liquor; rather, they were keener on imposing basic hygiene upon distillers and reducing instances of liquor-induced public disorder. Therefore, the emphasis on health and sanitation seems to be Bengal specific, probably due to the prevalent notion that cholera was a disease of locality and could not have emerged in full virulence other than in the Bengal delta.

Crime and Crimps in Sailor Dens

Surgeons in the Indian Medical Service suggested that consumers of intoxicants containing certain narcotics were more likely than others to commit a crime. It was commonly believed that respite from long and difficult sea voyages caused sailors to abandon every pretence of discipline, drink with abandon, and act insubordinately. The police frequently received complaints about sailors' drunkenness, assault, theft, refusal to work, absence without leave, inability to pay fines, suspicious loitering, rioting, and indecency (Cave-Browne 1865, 462–463). Several decades of court records indicate that many of the crimes committed by sailors were perpetrated under influence of adulterated liquor sold in the markets. There were also complaints that the government had not done enough to stop such illegal activities. Juries in Calcutta regularly pleaded to the court to prohibit this 'evil' on the rise in the streets and markets of the city. Nevertheless, despite the power invested in it, the first Sanitary Commission accomplished less than was anticipated (*The Friend of India*, 2 June 1864).

Some Europeans in Calcutta also emphasized that it was wrong to criticize all mariners as a 'drunken, reckless, mutinous lot', since it was the difficult circumstances of seafaring that drove them into disreputable activity (Cave-Browne 1865, 453). It is hard to determine the actual number of 'criminal' sailors, as police records are sparse, and numbers were often inflated. Malleson (1866) referred to a lock-up register that contained 365 instances of sailor drunkenness and confinement. Of these, 186 comprised mariners living ashore—nearly 3 per cent of the off-duty sailors. Many among them were charged with assault. However, only 35 were committed to the Sailors' Home for corrective measures. Repeated complaints about sailor behaviour maligned them in the eyes of some, but also sparked attempts by others to understand and redress their problems (Cave-Browne 1865, 462–464).

Additionally, Flag Street, the residential neighbourhood of sailors, facilitated contact with local street women and prostitutes for whom Chevers (1864, 65) recommended the construction of licensed and regulated lock hospitals—establishments that specialized in treating sexually transmitted diseases. Many argued that sailors could hardly avert the temptations offered in Flag Street, the Sailor's Home, or any boarding house in disreputable neighbourhoods. They needed some amusement to keep themselves busy. For example, Seamen's Chaplain A.L. Mitchell advocated the creation of an institute where sailors might socialize; entertain themselves with a large bowling alley, chess, and draughts; and drink tea, coffee, 'good' soda water, ginger beer, and lemonade at proper rate (Malleson 1866, x–xiii). His dream was realized when the Seamen's Reading and Coffee Rooms were opened under the auspices of the Methodist Church at 19 Lall Bazar in 1874. The institution conducted a religious service every night for seamen visiting Calcutta and maintained a coffee room that offered excellent refreshment at low prices and a reading hall with newspapers, magazines, and about 500 books gifted by 'friends of the institution'. The main agents of the institution were a group of women, including Mrs. Meik, Mrs. Conklin, and Mrs. Henderson, who visited grog shops and invited the seamen to the services (*Annual Report of the Missionary Society of the Methodist Episcopal Church* [henceforth ARMSMEC] 1892, 222). Rev. Frank W. Warne quoted George Henderson, the person responsible for the coffee rooms and seamen's work, saying in 1891 that sailors regularly attended and greatly appreciated the refreshment rooms at the mission (ARMSMEC 1891, 202).

The impetus behind opening the Sailors' Home in 1837, which fell into disrepute soon afterwards, was to 'suppress crimping and all the evils arising from it to which owners, commanders, officers and crews are subject in the port of Calcutta' (*Madras Missionary Register*, January 1838, 122–123). Sailors' Home in Bombay and Madras came up later in the same year. Mitchell gave a vivid description of how crimps operated in Calcutta in his narration of destitute white sailors. The 'crimping system' was practised widely on board ships, in Flag street, and in the vicinity of the shipping office. The members of the gang of crimps called themselves 'runners'. He accused them of being 'harpies' who enticed sailors to consume drugged liquor. The sources of Mitchell's information were convalescent sailors, who presented themselves as helpless, 'unfortunate victims' of crimping, as 'dupes of conspiracy' (Malleson 1866, xiii). These despondent sailors wandered around the city, often ending up in prisons. The problem persisted throughout the nineteenth century and, according to an observer, resulted in the sailors' debilitation 'from the effects of methylated spirits administered under the name of gin or whiskey, etc., by opium, or cocculus indicus in the name of beer, and in addition, by some loathsome complaint that may and probably will incapacitate him after a day or two at sea' (*The Nautical Magazine*, June 1871, 385).

Crimps were reported to be ubiquitous in the sailors' quarters of London—boarding houses, tap rooms of public houses, long rooms of gin palaces, and brothels—and to have ruined many a maritime career. An anonymous writer to the *Sailors' Magazine* (A Captain and His Mate 1845, 139–141) blamed sailors for landing themselves in traps they were well aware of and ship owners and captains for insensitively driving sailors towards destitution. The propensity to relax in the port city after too much hard work and lack of recreation aboard vessels led to drinking binges. An inebriated sailor could be easily persuaded to turn to crime and a 'libertine' life. The writer refers to 'foes' that drugged the drinks of seamen who risked becoming addicted and 'enslaved' to the 'virile' poison. By the time the sailor had recovered from his drug-induced reverie, these crimps would have disappeared with his belongings—clothes and money. Subsequently, unable to afford meals and accommodation, the sailor would have no choice but to depend on the same or another crimp for credit.

The crimps, masquerading as employment agents, carried placards reading 'able seamen wanted'. They advanced sailors credit for exorbitantly priced accommodation and clothes to be paid from any future earnings. Again, unscrupulous ship owners gave commissions to crimps who could provide mariner labour cheaply at short notice. A 'drugged', 'stupefied', distressed sailor was in no position to bargain for a proper wage. As one ship captain reported to the *Sailors' Magazine* (October 1844, 116–119), 'thus is the most noble and most generous of Britain's sons duped, before he sets his foot ashore'. The effectiveness of the Sailors' Home and the judiciary in protecting seamen from crimps was continually questioned. *The Calcutta Christian Observer* (September 1841, 590–591) wrote that the Calcutta Seamen's Friend Society should strive harder to uproot the crimping system. It argued that a comparison between the numbers of sailors provided jobs by the Sailors' Home and crimps, or sailors finding their way to the Home and to crimps, would illustrate the success of sailors' welfare measures. It suggested to the managers of the Home to seek police cooperation to dispose of crimps, ask shipping lines to give jobs only to seamen sent by them, and employ a number of agents for visiting ships on their arrival and shepherding sailors to the Home. Evidently, some people, Christian missionaries in particular, considered crimping a threat to the physical and moral wellbeing of sailors.

Conclusion

This chapter examines the significance of the social and moral concern for seamen visiting Calcutta to colonial medical and sanitary policy. It discusses the issue of the propensity for drunkenness among sailors as manifested in its impact on their health and demeanour. The efforts to alleviate cholera and adulteration of liquor provide important insights into the ambivalence of the early colonial administration in India. The colonial state devised policies in response to the threat Indian society was considered to pose to white sailors, who were variously labelled as disruptive and mostly vulnerable, and thus destabilized the carefully constructed idea of imperial British identity consisting in righteousness and discipline. The transfer of medical knowledge from Britain to India, and its translation into public policy, formed a salient feature of imperial formation, designed to protect the racial superiority and honour of colonizers from the threats posed by both white subalterns and Indians.

Despite similarities in the concern for seamen's health in British and Indian ports, the methods of addressing them were not uniform. Europeans tended to describe their sailors outside Europe sympathetically, usually as innocent and sober men struggling with a challenging environment. Thus Calcutta was described as an unhygienic port city whose 'natives' were responsible for many of the problems experienced by visiting sailors. The Europeans and Americans who made up the bulk of the city's crimp population and tavern owners were not criticized as vigorously as were Indians who sold poor quality liquor. Indian crimps were portrayed as conspirators who tricked 'innocent' European sailors into consuming low-quality liquor, sold often in bottles bearing labels of English brands that caught them unawares. Despite crimps across continents operating in similar fashions, British commentaries on Indian crimps suggest a racial differentiation. In sum, this chapter explores the living conditions of European sailors visiting India and the changing image of such sailors in the public and official mind.

References

A Captain and His Mate. 1845. The Crimping Sytem in London. *The Sailor's Magazine 17* (5, January): 139–141.

Accum, Fredrick. 1820. *A Treatise on Adulteration of Food and Culinary Poisons, Exhibiting The Fraudulent Sophistications of Bread, Beer, Wine, Spirituous Liquor, Tea, Coffee… and Methods of Detecting Them*. London: Longman.

Annual Report of the Missionary Society of the Methodist Episcopal Church for the Year 1891.

Annual Report of the Missionary Society of the Methodist Episcopal Church for the Year 1892.

Annual Report on the Administration of the Bengal Presidency, 1864–1865.

Anonymous. 1992. Grog Shops. In *Calcutta Through British Eyes 1690–1900*, ed. Laura Sykes, 47–48. Madras: Oxford University Press.

Arnold, David. 1979. European Orphans and Vagrants in India in the Nineteenth Century. *The Journal of Imperial and Commonwealth History* 7: 104–127.

———. 1991. The Indian Ocean as a Disease Zone, 1500–1950. *South Asia: Journal of South Asian Studies* 14 (2): 1–21.

———. 1993. *Colonising the Body: State Medicine and Epidemic Disease in Nineteenth-Century India*. Berkeley: University of California Press.

Barrett, T.R. 2004. *Calcutta: Strange Memoirs – Foreign Perceptions*. Calcutta: Deep Prakashan.

Cave-Browne, John. 1865. Sailor Life in Calcutta. *Calcutta Review* 40: 452–466.

Chakrabarty, Dipesh. 1991. Open Space/Public Place: Garbage, Modernity and India. *South Asia* 14: 15–31.

Chevers, Norman. 1864. *On the Preservation of the Health of Seamen, Especially Those Frequenting Calcutta and Other Indian Ports*. Calcutta: Military Orphan Press.

———. 1886. *A Commentary on the Diseases of India*. London: J. & A. Churchill.

Conley, Mary A. 2009. *From Jack Tar to Union Jack: Representing Naval Manhood in the British Empire, 1870–1918*. Manchester: Manchester University Press.

Firminger, Walter K., ed. 1906. *Thacker's Guide to Calcutta*. Calcutta: Thacker, Spink.

Fischer-Tiné, Harald. 2012. The Drinking Habits of Our Countrymen: European Alcohol Consumption and Colonial Power in British India. *The Journal of Imperial and Commonwealth History* 40: 383–408.

Gastrell, J.E., and H.F. Blandford. 1865. *Report on the Calcutta Cyclone of the 5th of October, 1864*. Calcutta: Military Orphan Press.

Gray, Taylor. 2016. Whispers in a Bottle: The Language of Liquor and the Politics of Prohibition in Nineteenth-Century New Zealand. *Law and Humanities* 10: 90–114.

Gustafson, Axel. 1884. *The Foundation of Death: A Study of the Drink-Question*. London: Kegan Paul.

Harrison, Mark. 1999. *Climates and Constitutions: Health, Race, Environment and British Imperialism in India 1600–1850*. New York: Oxford University Press.

Hassall, Arthur H. 1861. *Adulterations Detected or Plain Instructions for the Discovery of Frauds in Food and Medicine*. 2nd ed. London: Longman, Green, Longman, and Roberts.

Haughton, James. 1855. *A Plea for Teetotalism*. London: William Tweedie.

Joyce, Michael K. 1854. *An Exposure of the Haunts of Infamy and Dens of Vice in Bombay*. Bombay: Bombay Gazette Press.

Macpherson, Hugh M. 1861. *On the Mortality of Calcutta During the Twenty Years Ending with 1860*. Calcutta: s.n.

Malleson. G.B. 1866. *The State of Sailors in Calcutta*. Marine Proceedings May 1866. British Library IOR/P/433/52:1866.

Marks, Shula. 1997. What Is Colonial about Colonial Medicine? And What Has Happened to Imperialism and Health? *Social History of Medicine* 10: 205–219.

Massey, Montague. 1918. *Recollections of Calcutta for over Half a Century.* Calcutta: Thacker, Spink.

Morton, Sandra. 2005. A Little of What You Fancy Does You… Harm!! (with Apologies to Marie Lloyd). In *Criminal Conversations: Victorian Crimes, Social Panic, and Moral Outrage*, ed. Judith Rowbotham and Kim Stevenson, 157–178. Ohio: The Ohio State University.

Murdoch, John, ed. 1863. *India Year-Book for 1862.* Madras: Graves, Cookson & Co.

Otter, Chris. 2006. The Vital City: Public Analysis, Dairies and Slaughterhouses in Nineteenth-Century Britain. *Cultural Geographies* 13: 517–537.

Peers, Douglas M. 1998. Soldiers, Surgeons and the Campaigns to Combat Sexually Transmitted Diseases in Colonial India, 1805–1860. *Medical History* 42: 137–160.

———. 1999. Imperial Vice: Sex, Drink and the Health of British Troops in North Indian Cantonments, 1800–1858. In *Guardians of Empire: The Armed Forces of the Colonial Powers c.1700–1964*, ed. David Killingray and David Omissi, 25–52. Manchester and New York: Manchester University Press.

Preston, Theodore J. 1895. Epidemic Disease in the Royal Navy. *Transactions of the Epidemiological Society of London* 14: 31–68.

Tambe, Ashwini. 2009. *Codes of Misconduct: Regulating Prostitution in Late Colonial Bombay.* Minneapolis and London: University of Minnesota Press.

Wald, Erica. 2012. Health, Discipline and Appropriate Behaviour: The Body of the Soldier and the Space of the Cantonment. *Modern Asian Studies* 46: 815–856.

CHAPTER 9

Chikungunya and Epidemic Disease in the Indian Ocean World

Edward A. Alpers

Introduction

Twenty-five years ago, David Arnold (1991) published a pioneering article on the medical history of the Indian Ocean. In his wide-ranging essay, he emphasized the impact of European intrusion into the Indian Ocean world (IOW) after 1500, noting especially that "the emergence of India as the lynch pin of British power and trade in the East was of great epidemiological significance for the rest of the region and indeed the wider world beyond" (ibid., 7). Arnold particularly acknowledged the importance of maritime movement, whether by traders, soldiers, pilgrims or migrant labourers, as a central element in disease dispersion—what he called "epidemiological routes and conjunctures"—during the post-contact period (ibid., 9). More recently, Amina Issa (2006) has built on Arnold's essay by arguing for the significance of indigenous sailing ships and the diffusion of epidemic disease in Indian Ocean ports in the nineteenth century before the advent of steamships. In this chapter, I propose to jump forward in

E. A. Alpers (✉)
University of California, Los Angeles, Los Angeles, CA, USA
e-mail: alpers@history.ucla.edu

G. Campbell, E.-M. Knoll (eds.), *Disease Dispersion and Impact in the Indian Ocean World*, Palgrave Series in Indian Ocean World Studies, https://doi.org/10.1007/978-3-030-36264-5_9

time to the contemporary period by examining the chikungunya epidemic that swept across the IOW in the first decade of the twenty-first century, and to ask whether it is possible to apply what we have learned from studying the origins and spread of this disease to improving our understanding of earlier epidemic diseases in the IOW.

Although my background is not in the history of medicine, I have long been familiar with the remarkable work of Dr James Christie, who served as physician to the Sultan of Zanzibar for a decade from 1865 (Anon. 1892). My primary interest in his account of cholera epidemics in East Africa (Christie 1876) was how vividly they revealed caravan trade routes linking the Swahili coast and the Nile Valley to the interior. In his very first contribution on cholera in East Africa, Christie observed: "In all the turnings and windings of the cholera epidemic, there seems to have been one unvarying principle directing its course. It has invariably accompanied trade wherever its direction may have been" (Christie 1871, 115).[1] Here, however, I look more carefully at Christie's medical work as an epidemiologist. This re-examination prompts me to use it as a springboard for thinking about disease diffusion in general and the difficulty of identifying diseases from the historical record.

Chikungunya as a Way to Think About Epidemic Disease

So where does chikungunya enter this picture, and how can an analysis of this latest example of epidemic disease help historians to think about earlier examples of disease dispersion in the Indian Ocean world? As I hope to demonstrate, it is the genetic ability of biological organisms to adapt to changing circumstances—whether natural or anthropogenic—through mutation that is the key lesson from the history of chikungunya. The chikungunya virus was not firmly identified scientifically until an outbreak occurred in 1952, in the southeast of present-day mainland Tanzania, when strains of a new virus were identified from the sera of afflicted patients (Robinson 1955; Ross 1956). This newly identified virus was given the local Kimakonde name "chikungunya", meaning "that which bends up", because of the clinical symptoms of "the severe arthralgia

[1] For modern citations of Christie's work, see Hartwig (1975, 63), Koponen (1988, 159–162, 173–176), Kjekshus (1996, 23–24) and Rockel (2006, 132), and the critical commentary by Echenberg (2011, 53).

[joint pain] that is a hallmark of chikungunya fever, the disease caused by the virus" (Weaver and Lecuit 2015, 1231). The chikungunya virus (CHIKV) is an RNA virus that belongs to the family *Togaviridae*, genus *Alphavirus*, and is part of the Semliki Forest virus antigenic complex (Sergon et al. 2007, 1189). There are three major genotype lineages of CHIKV: East/Central/South African (ECSA), the original Tanzanian strain; West African and Asian. Identification of the new virus represented a major breakthrough in the analysis of mosquito-borne tropical fevers since it was initially described as a form of "dengue" fever, a designation that for many years had served as a broad lumping category for a wide variety of clinically similar diseases. When Donald Carey began to look into the fever's history in the early 1970s, he discovered that what had in many instances been called "dengue" may in fact have been chikungunya. Most interestingly, exploring the history of chikungunya leads us back to James Christie, who, as Carey comments, penned "an important and fascinating epidemiologic account of dengue" in 1881 (Carey 1971, 255).[2]

After he left Zanzibar in 1874 and returned to Glasgow to practice medicine and publish his *magnum opus* on cholera, Christie continued to draw upon his East African experiences to indulge his interest in tropical epidemic diseases. His paper "On Epidemics of Dengue Fever: Their Diffusion and Etiology" (Christie 1881), cited by Carey, was a product of that continuing curiosity, but Carey overlooked a paper that Christie wrote a decade earlier while still based in Zanzibar, which, like his writings on cholera, was based on his exceptional powers of observation. Christie reported that in July 1870, "after the complete disappearance of cholera from the island of Zanzibar, a new form of fever, quite unknown to the bulk of the population, was epidemic in the island of Zanzibar, and more especially in the town, where it attacked almost the entire population" (Christie 1872, 21). In 1871, Christie recognized its similarity to descriptions of dengue by other physicians in other locations, but he also saw that this fever had certain different characteristics. When he returned to this subject a decade later, he explained that "the older inhabitants recognized the disease as one which had been epidemic about forty-eight or forty-nine years before, and they gave to it its former designation—Ki-dinga Pepo" (Christie 1881, 165; also quoted by Carey 1971, 255), meaning "cramp-like pains, caused by an evil spirit" (Christie 1872, 22). Christie

[2] Carey had studied endemic dengue in south India in the 1960s; when he published this article, he had a position at the Rockefeller Foundation of the University of Ibadan, Nigeria.

made several astute observations that have been confirmed by methods unknown in the nineteenth century. First, he observed: "Regarding the epidemic of 1870, I may state that it was not introduced to Zanzibar from without. The disease appeared at the height of the south-west monsoon, and for at least three months before, dhow communication from the north was impossible" (Christie 1881, 172). In view of his understanding of the spread of cholera, it should come as no surprise that he also noted: "With regard to the diffusion of the epidemic from Zanzibar to the mainland, I can only state generally that it took place along the lines of human intercourse, and that the coast towns were infected" (ibid., 166). What I find most remarkable, however, is Christie's hypothesis concerning the aetiology of *kidinga pepo*, which he speculated might possibly have been linked to a transformed version of cholera. "If chemical changes, of such a nature [human decomposition], take place in the *cadaver* [*sic*], may not physiological or pathological changes also take place? The hypothesis is not unscientific". Inspired by the work of Charles Darwin, he wondered further: "If the germ theory of diseases be admitted, the possibility of hybridization must be admitted also" (ibid., 174–175). Since Christie's conditional statements were penned before the great discoveries in virology, at a time when the miasma theory of contagious diseases was still dominant, this is all the more remarkable (Karamanou et al. 2012). The history of this century's chikungunya pandemic appears to bear him out.

What led me to take chikungunya as a starting point for thinking about the keynote presentation on which this chapter is based was the combination of the terrible impact it had on La Réunion in 2005–2006, where more than a third of the total population was affected; the disease's incredibly rapid and extensive diffusion across the IOW in the last decade and the ability of medical science, armed with the tools of genetic analysis, to track its aetiology definitively (see Her et al. 2009, 1165–1166). At the outbreak of the chikungunya epidemic, when in 2006 an editorial in a major French medical journal admitted that the chikungunya epidemic was completely "unexpected", virtually no scientific attention was being paid to this arbovirus. "Up to then unknown to both the public and to most physicians, a virus with an unpronounceable name, Chikungunya (CHIK), has invaded French news for several months" (Simon et al. 2006, 437).[3]

[3] See also early reports in Paganin et al. (2006); Pierre et al. (2006); Josseran et al. (2006).

Even at this early stage in the medical alert, it was recognized that the origin of the virus was in East Africa, that it had spread to the Comoro Islands before reaching Réunion and that "the diffusion among the islands had been promoted by the intensity of their human exchanges and their proximity (from several hours to a few days by boat)" (Simon et al. 2006, 439). What makes the history of the chikungunya epidemic so remarkable, however, are the genetic changes in the virus itself that enabled chikungunya to wreak such devastation on Réunion and eventually in India, where at least 1.4 million cases were recorded during the same 2005–2006 epidemic.[4]

It is now firmly established that CHIKV originated in Africa and that there exist two lineages of the virus, one West African and the other East/Central/South African [ECSA] (Burt et al. 2012, 662). There is also a genetically distinct Asian genotype that derived from ECSA in the late nineteenth or first half of the twentieth century (Weaver and Forrester 2015, 35). According to Felicity Burt et al., "In Africa, the virus is maintained in a sylvatic transmission cycle" between *Aedes* mosquitoes, which are "the main vectors", and small primates, mostly monkeys, although birds, cattle and rodents are also hosts (as distinct from dengue [DENV], where only primates act as hosts). Among mosquitoes, the most widespread vector is *A. aegypti*. An important distinction, however, is that CHIKV may be exchanged directly between humans and mosquitoes without animal intermediaries during epidemics (Burt et al. 2012, 662). Kamran Khan et al. (2014, 3, 15) state that humans infected by the virus soon become viremic and can transmit the virus directly to these insect vectors. In May 2004, a virulent outbreak occurred on Lamu Island, Kenya, where it had never previously been recognized. A serosurvey revealed an attack rate of 75 per cent in a total population of about 18,000 (Sergon et al. 2008, 335). The epidemic also struck Mombasa and by January 2005 had reached Ngazidja (Grande Comore), where, through 5 May 2005, 202 cases were reported. The same research team that investigated the Lamu outbreak was asked by the Comorian authorities to conduct a seroprevalence study on Ngazidja, where they identified a 63 per cent infection or seropositivity rate (Sergon et al. 2007, 1191). Taken together, these surveys suggest that many thousands of individuals on both Lamu and Ngazidja were infected by chikungunya.

[4] For an early report on the Indian epidemic, see Lahariya and Pradhan (2006).

Although it is not surprising that CHIKV spread from coastal Kenya to the Comoros in view of the long and intimate history connecting Ngazidja to Unguja, the main island of Zanzibar, it is noteworthy that the epidemic did not register significantly on Zanzibar. This may reflect limited research on Zanzibar, as in all of Tanzania, or the clinical confusion of CHIKV with malaria (Kajeguka et al. 2016),[5] but a small serosurvey of dengue (DENV) at Chake Chake Hospital on Pemba Island indicated that, while 15.4 per cent of patients tested revealed seroprevalence for dengue, none evidenced chikungunya (Vairo et al. 2012, e45). Without yet understanding the island-hopping nature of the CHIKV from coastal Kenya directly to Ngazidja, and having no direct evidence, it seems probable that the virus was transmitted by human carriers flying on the regular Kenya Airways flight from Mombasa to Moroni, the capital of Ngazidja. From Ngazidja, the virus spread to Mayotte, Mauritius, the Seychelles and Réunion, where, as we shall see, it had a major impact, precipitating a flurry of scientific research both to determine the cause of the outbreak and to recommend preventive measures, a topic addressed in this volume in the chapter by Karine Aasgaard Jansen and elsewhere (Jansen 2016). Indeed, increased air travel is one important factor generally recognized in the literature in promoting the rapid and widespread distribution of CHIKV (Renault et al. 2012; Morrison et al. 2016).[6]

The fundamental question is what precipitated the CHIKV to re-emerge and cause such damage after decades in retreat. For although there were outbreaks of the Asian strain of chikungunya in Southeast Asia at the end of the twentieth and beginning of the twenty-first centuries, the virus has been reported only occasionally in Africa since the 1960s (Lam et al. 2001; Laras et al. 2005; Burt et al. 2012, 662–663). Here the answer lies in a single genetic mutation in CHIKV. Genetic analysis was based on viral isolates taken from 127 patients from all the affected southwest Indian Ocean islands, plus Madagascar. The authors of this important study (Schuffenecker et al. 2006) traced the cause of the epidemic to evolution in the East African strain. Specifically, they discovered that "the emergence of genotype E1-226V, which was observed from the beginning of September 2005 and experienced a spectacular rise in frequency…pre-

[5] Kajeguka et al. (2016, 7/9) conclude: "Chikungunya virus appears to be actively circulating in the population", which is not surprising given its endemic character in Africa.

[6] More generally, Siddhartha Mukherjee (2016, 33) notes that "viruses prefer to travel these days—on transcontinental airplanes".

ceded the explosive epidemic peak of mid-December 2005". This suggested to them "that such a mutation provides a selective advantage to the virus in mosquitoes". When they published their report in July 2006, they considered that the epidemic began in the Comoros, since the first verified cases reported in March 2005 on Réunion were linked to travellers from those islands, undoubtedly from Mayotte, which, like Réunion, is an overseas French territory (ibid., 1059). However, a subsequent study suggested that the ultimate source of the virus could be located in coastal Kenya, since the genetic footprint was the same as in Réunion (Njenga et al. 2008, 2757). Its authors also surmised that further mutations may have occurred and exacerbated the epidemic, a scenario at which Schuffenecker and her colleagues also hinted.

Yet, despite its important contribution, the authors did not fully recognize what it was, specifically, about this mutation that rendered it so effective in spreading CHIKV on Réunion. This discovery was made by a research team from the University of Texas Medical Branch at Galveston, which discovered

> that a single nucleotide change, which arose during the epidemic, significantly increases fitness of the virus for *Ae. albopictus* mosquitoes and was associated with CHIKV dependence on cholesterol in the mosquito cell membrane. This change likely enhanced CHIKV transmission by an atypical vector and contributed to the maintenance and scale of the epidemic. (Tsetsarkin et al. 2007, 1896)

The key here is that, whereas the vector for CHIKV in East Africa and the Comoros was the familiar *A. aegypti*, which was absent on Réunion, "an E1-226V mutation in CHIKV results in increased fitness of CHIKV in *Ae. albopictus* mosquitoes", the much hardier Asian tiger mosquito (so-called because of the stripes on its middle body section), which was endemic on that island (ibid., 1900). By itself, it appears that "a single mutation is sufficient to modify viral infectivity for a specific vector species and as a consequence, can fuel an epidemic in a region that lacks the typical vector" (ibid., 1901). Furthermore, the team argue that, prior to this mutation, the level of CHIKV viruses in the blood was sufficient to indicate *A. albopictus* mosquitoes as the source (ibid.). In other words, when the initial Comorian travellers entered Réunion carrying CHIKV, they served as a human reservoir that, once bitten, was able to transform the hitherto unaffected *A. albopictus* into a vector for CHIKV. In addition, the team contends, this mutation gave the Asian tiger mosquito a selective advantage over the African mosquito that was the fever's primary vector in Africa. Accordingly, once this rapid adaptation occurred in *A. albopictus*, it accelerated CHIKV so that it rapidly infected what was an entirely unprotected human population (ibid., 2903).[7] One particular advantage of the *A. albopictus* vector is its great adaptability in an urban environment, where it had evolved in Asia (Pialoux et al. 2006, 255).[8] In the end, the attack rate on Réunion was somewhere between 34.3 per cent and 38.2 per cent, depending on the mode of analysis, in a population of more than 700,000, yielding estimates ranging from 244,000 to 266,000 to as high as 300,000 cases of chikungunya. Moreover, for the first time ever, some 203 deaths were reported that were attributed to the virus (Renault et al. 2007; Pialoux et al. 2007; Gérardin et al. 2008).

Nevertheless, a few years after the chikungunya outbreak, Njenga et al. (2008, 2758) lamented that there was still no satisfactory explanation for the cause of such a pandemic. Notwithstanding the identification of the E1-226V mutation and its ability to transform *A. albopictus* into a vector for CHIKV, it was evident that the major changes in CHIKV had occurred before its earliest known appearance on Lamu in 2004. Moreover, while prior outbreaks in Africa were caused by strains within the same ECSA genotype, none were apparently ancestors of these new strains, so that their origins might have been any of a number of existing strains in either Africa or Asia (ibid., 2579). A year later, however, impelled by the extension of the pandemic to India, where it affected 13 states and at a minimum more than a million people, by applying several different modelling systems, researchers from the National Institute of Virology in Pune, Maharashtra, had succeeded in identifying a Ugandan strain that they could date to about 2000 (Cherian et al. 2009). Thus, within a few years of this completely unanticipated CHIKV epidemic, which had spread across the entire IOW, the Singapore-based authors of a modelled network for this historical process concluded that "current knowledge underscores the complexity of the vector-virus-environment interactions, and clearly demonstrates their role in changing the infectious disease epidemiology" (Ng and Hapuarachchige 2010, 882). Even though in 2013 much was yet to be discovered about how the E1-226V variant emerged,

[7] For a simplified synopsis, see Anon. (2007).

[8] For a recent example of how this mosquito and this disease have invaded a new urban territory, see Kampango and Abílio (2016), Gudo et al. (2015).

the authors of a validating follow-up report emphasize the broader point that arboviruses have the ability to invade a new host as a result of the process of genome replication (Arias-Goeta et al. 2013, 1). I should note, as well, that the structure of the Asian strain prevents it from acquiring the E1-226V variant unless another substitution, E1-T98A, is also present (Weaver and Forrester 2015, 35, Box 1). It is appropriate that the genomic strain created by the E1-226V mutation identified by this global research into the CHIKV pandemic has been named the Indian Ocean Lineage (IOL).

A further complication to the history of CHIKV is that its main vector in the south Indian outbreak of 2005–2006 was not the Asian tiger mosquito, but *A. aegypti*; nor was the strain of the virus the older Asian genotype, but the ECSA strain (isolated at Yawat, Maharashtra) that we now know carried the E1-226V genotype associated with the radical transformation of the virus first observed at Lamu and Mombasa (Yergolkar et al. 2006). Aggravating the situation in India, however, was a "second-step mutation" that appears to have made *A. albopictus* an even more efficient vector for CHIKV in other parts of India, where between 2006 and 2011, CHIKV infection spread to 19 states and affected as many as several million individuals (Tsetsarkin and Weaver 2011, 1). According to the Galveston group of medical scientists, they suggest that a "novel substitution, E2-L210Q, identified in Kerala, India in 2009, caused a significant increase in the ability of CHIKV to infect and develop a disseminated infection in *A. albopictus*" (ibid., 2, Author Summary). In fact, subsequent research by this same team indicates that this substitution is only one of four such mutations that cause even greater increases in infectivity (Tsetsarkin et al. 2016, Table 1). Stated in more technical terms, they find that

> adaptation of CHIKV to a new mosquito vector can be a multistep process that, since 2005, has involved at least 2 amino acid substitutions in the envelope glycoproteins. The substitution that provides the strongest selective advantage, E1-226V, was followed by second adaptive mutation (E2-L210Q) that has resulted in a strain circulating in India with the fittest phenotype detected yet for transmission by *A. albopictus*. (ibid., 11)

Further evidence of the remarkable ability of CHIKV to adapt rapidly to new circumstances may be found in the identification of an entirely novel CHIKV outbreak strain among 19 mutations in India that did not contain the E1-A226V mutation (Kumar et al. 2014). The authors of a subsequent study into this phenomenon write that

> after CHIKV reached the first-step E1-226V *A. albopictus*-adaptive peak, its evolution was no longer constrained to a monolithic peak and multiple adaptive peaks of relatively equal fitness became available for Darwinian evolution. (Tsetsarkin et al. 2014, 10)

They continue, in language that recalls Christie's allusion to Darwinian evolution:

> Overall, our findings somewhat mirror traditional Darwinian models of macroevolution, where major adaptations, such as development of wings by ancestors of birds, or the E1-226V substitution in the case of CHIKV, can result in the rapid radiation/diversification of new lineages/species. (ibid.)

These findings imply that the combination of a bundle of environmental factors, including climate change and population growth, together with increased urbanization, may give rise to future arbovirus outbreaks in the Indian Ocean world (Gaüzère et al. 2012; Abstract).[9]

So was Carey (1971, 261) correct when he concluded of diseases clinically reported in Java in 1779, Zanzibar in 1823 and 1870 (by Christie), and India in 1824, 1871, 1902, 1923 and 1963–1964 that all appear to have been chikungunya or something akin to it? Perhaps. But in the absence of the kind of genetic evidence that we possess for the 2005–2006 outbreak that riveted the attention of modern medical researchers, I tend to concur with the more cautious conclusion of Goro Kuno (2015), who recently reviewed the same confusion between dengue and chikungunya that Carey observed 35 years ago. Kuno cautions that the absence of available details about past diseases, changing diagnostic methods and the descriptive language of observed symptoms render after-the-fact historical identification of diseases extremely difficult. Accordingly, it remains an open question as to whether what in the past has been identified as dengue may have been chikungunya.

[9] For a discussion of "the current geographic range and the relevant biological traits of *A. albopictus* in order to explain its rapid spread", see Paupy et al. (2009); also Delatte et al. (2011).

Comparing Chikungunya to Malaria in the Mascarenes

Notwithstanding Kuno's cautionary words, I propose to take these lessons from the very recent history of chikungunya detection and apply them to our thinking about the longer-term history of disease dispersion in the IOW. I am under no illusions about the limitations of such an exercise, but I would like to assess what may be possible in the following section of my chapter. To do this, I will focus on the disease history of Mauritius and Réunion. These two remote islands share a deep history of mid-ocean isolation and the absence of indigenous human populations before European colonization, as well as closely linked colonial and settlement histories. They also shared many of the same epidemic diseases that reached them across the water during the first centuries of French and, later, British colonial rule.

Historically, perhaps the most interesting comparison with chikungunya is malignant malaria, which struck Mauritius in 1867 as an endemic crowd disease and then Réunion in 1869.[10] Quite apart from early (and some later) depictions of the Mascarenes as Edenic in their healthfulness (compare Ève 2009), there is no evidence that this most deadly form of malaria existed endemically prior to its introduction in the 1860s. However idealized this image may have been, especially in light of the history of both smallpox and cholera epidemics in the decades preceding the malaria outbreak (Boodhoo 2010, 57–62, 134–137, 143–145, 149, 159), it is quite different from the negative reputation of the notoriously unhealthy Maldive Islands (Knoll, this volume and 2018). In his 1908 study of malaria prevention in Mauritius, Nobel Laureate Sir Ronald Ross, who discovered the link between the malaria parasite and its mosquito-borne vector, argued that, although there were undoubtedly some, probably many, captive Africans and Malagasies, as well as Indian indentured labourers, who came to the island carrying malaria, it was not present there as an endemic disease before the 1860s (Ross 1908, 44). In his opinion, but begging the question of what is meant specifically by malaria, had malaria been present even minimally, people in Mauritius would have recognized it and commented on its presence (ibid., 45). The fact that the island was

[10] From a contemporary public health perspective, both dengue (DENV) and the zika virus are clearly important comparisons, especially as they relate to tropical islands like the Mascarenes. See Cao-Lormeau (2016).

used by the British as a hospital for ailing military personnel from its Indian Ocean empire holdings apparently convinced Ross that malaria was not a significant presence in Mauritius. To demonstrate this conclusion, he cites an analysis of medical statistics for British troop mortality rates in Mauritius, which show that upticks in malaria cases among British troops who had arrived on the island from India and China in the decade before 1867 were a result of relapse, rather than of the presence of endemic malaria (ibid.).

Like the challenge of identifying chikungunya noted above, Ross points to the presence of other fevers in Mauritius before the malaria epidemic of 1867 that complicated matters of disease identification (ibid., 46). Most significant among these was a form of epidemic relapsing fever that mainly affected the Indian population and was known locally as "Bombay fever" (ibid., 47). However, when the malaria epidemic of 1867 hit Mauritius, for Ross there was no mistaking the fact that it represented an entirely different disease. As Ross noted, "Accounts of eye-witnesses of the fever at Port Louis recall descriptions of plague and cholera" (ibid., 48). More than one-fifth of the population of Port Louis perished from the fever in 1867, while almost 9 per cent of the total island population died, and in the words of the Fever Enquiry Commissioner's Report of 1868, "the survivors 'were so prostrated by disease that the living were scarcely able to bury the dead'" (ibid.).[11] What Ross clearly described as malaria was unquestionably *Plasmodium falciparum*; what he dismissed as not being malaria may, however, have been *Plasmodium vivax*, *P. ovale* or *P. malariae*, the three other most common species of this protozoan parasite that infect human beings.

"This astonishing occurrence", Ross wrote, "caused much perplexity at the time among the more thoughtful students of malaria. It showed that the disease is at all events not due to any inherent poisonous property of soil, but rather that it might be caused by some living organism capable of invading a country from without". Indian coolies were the favourite target for such suspicions, as were visiting ships, as well as various natural phenomena, such as cyclones. But as we know, and as Ross argued, a solution

[11] Here I must note the very great significance of the inauguration of the combined archival and historical archaeological research at the Bois Marchand Cemetery in Mauritius, a burial place specifically opened to deal with the exceptional number of deaths caused by the malaria epidemic of 1866–1867, being led by Krish Seetah of Stanford University and recounted to us at the 2016 conference (see British Library 2016).

lay in the not yet discovered identification of the mosquito as the disease vector (Ross 1908, 49). Ross considered two competing hypotheses, one that the appropriate mosquito had only recently been introduced to Mauritius, the other that it was already present, but needed a critical mass of infected carriers to become endemic. He preferred the first hypothesis, especially because of the almost simultaneous introduction of malaria to Réunion, and therefore attributed the 1867 outbreak to the recent arrival of a mosquito that he named *Pyretophorous costalis*, a subspecies of *Anopheles gambiae*, as the vector for the malaria parasite (ibid., 49–52).

Indeed, Ross's preference for the first hypothesis has become the accepted interpretation; but if the second were preferred, might it not be possible that, in light of the mutations that I have described above for CHIKV and its mosquito vectors, the same possibility exists for the history of malaria in Mauritius? What I am suggesting here is that, since we know that non-endemic malaria did in fact exist in Mauritius before 1867 in the bodies of enslaved Africans and Malagasies, recuperating British soldiers and Indian indentured labourers, is it not feasible that a similar genetic mutation might have occurred that transformed existing mosquito populations on the island so that they were in a position to become effective vectors for malaria? Certainly, Raj Boodhoo's careful presentation of contemporary efforts to understand "the malaria scourge", as he calls it, provides plenty of evidence for questioning Ross's conviction that earlier forms of fever in Mauritius were not malaria (Boodhoo 2010, 173–191). The tendency to blame its introduction on Indian immigrants, who were massively transforming the demography of the island colony, undoubtedly obscured the possibility that other varieties of malaria might have been introduced previously in non-endemic, non-malignant forms by both enslaved Malagasies and Africans. I find it interesting, for example, that quinine, the most widely used prophylaxis for malaria, was first used to treat fever patients in Mauritius as early as 1828 and that one local physician observed in 1868 that "intermittent fever had always existed" (ibid., 181, 184). It is worth adding here what William Twining noted of Bengal in 1832: "malaria has been generally acknowledged the efficient cause of intermittent fevers" (quoted in Mukherjee 2008, 55). Similarly, yet another medical doctor wrote of Mauritius in the late nineteenth century that "the island had never been a healthy country" (Boodhoo 2010, 186).

Absent from discussions of the origin of endemic malaria in Mauritius is any consideration of contemporary changes in the virulence of malaria in India, especially in Bengal, where "a savage new malaria was devastating regions previously healthy or lightly afflicted by the malady" (Klein 2001, 147). Specifically, the middle of the nineteenth century witnessed the advance of what was described as "malignant malaria" caused by *Plasmodium falciparum*, the deadliest form of the four species of the Plasmodium parasite that cause malaria in humans. Recalling Mauritian references to "Bombay fever", this newly virulent form of malaria was known by contemporary observers in India as "Burdwan fever" after the area of West Bengal from which it spread across the wider Bengal region; in other areas of Bengal, it was called "Jessor fever", "Nadia fever" or "Hughly fever" (ibid., 161). David Arnold (1999, 136) describes this disease as "the Bengali black death", while Ira Klein (2001, 159) argues that this deadly transformation in malaria was a consequence of the extensive ecological interventions that resulted from British development policies of railway, canal and road construction that transformed colonial India during these decades through deforestation and the increased prevalence of standing water. Indeed, these changes stimulated the proliferation of several different *Anopheles* mosquito vectors that carried *Plasmodium falciparum*. It seems almost certain that it was this "Burdwan fever" that was carried from Calcutta to Mauritius in 1865, a year in which more than 16,000 Indian emigrants embarked from that port aboard 43 ships (Deerpalsingh and Carter 1996, 312). Indeed, Arabinda Samanta (2002, 58) specifically comments on the fact that malaria moved to Mauritius in the mid-1860s during a period of heightened epidemic malaria in Bengal in her important monograph on epidemic malaria in Bengal. We have seen that Ross put forward only two hypotheses for the origin of malaria in Mauritius, either that the mosquito vector had only recently been introduced into Mauritius, or that it was present but required a sufficient number of infected carriers to become endemic. Might not this vivid example of Indian Ocean disease dispersion admit a possible third hypothesis, one that allows for the pre-existing presence of both mosquitoes and non-malignant forms of malaria in Mauritius *and* for the introduction of a newly virulent form of malaria plasmodium from Bengal? According to this hypothesis, these two forms would have combined in ways unknown to precipitate the epidemic of malignant malaria that devastated the island in the mid-1860s.

In the late nineteenth century, however, the question remained unresolved. When Dr Daniel Anderson visited Mauritius in 1889, he was struck by the fact that no one could yet agree on the origin of malaria on the island (Anderson 1918, 173; compare Floate 1969).[12] Yet he joined Ross in dismissing the earlier school of thought regarding unhealthy soils and claimed that it was not until the arrival of an immigrant ship named the *Spunky* from India in 1865 that malaria was introduced (ibid., 173, 175).[13] On the face of it, this assertion was quite inaccurate, as the British medical records examined by Ross reveal. In any case, at the time of Anderson's visit, the Anopheline mosquito had not yet been identified as the malarial parasite vector. However, in light of what was then known about the disease in West Africa, Anderson asked: "Are there any Anophelines in Mauritius and did they always exist there?" He confirmed that *Pyretophorous costalis* were indeed numerous, mainly in the low-lying coastal zone, but that no one had any idea of how long they had been established on the island (ibid.). Returning to the arrival of the *Spunky* and its malarial passengers, he explained that "Ross's mosquito-carrier discovery had not yet astonished the world". Thus, while his initial impression about how the parasite was somehow a consequence of the unhealthy air and water of the marshes around Port Louis proved to be wrong, the link between mosquitoes and that aquatic environment was correct (ibid., 176). Accordingly, addressing the London Hygienic Congress in 1890, Anderson could declare:

> Now we can answer our question in full. The two or three hundred malarial coolies landed at marshy Petite Rivière infected Anophelines, which at that time of the year and under the special circumstances that had favoured their extensive propagation were ready to bite the newly arrived infected immigrants, and to carry the parasite from village to village and estate to estate. (ibid.)

Staying with Anderson, he next shifted his emphasis from how and when malaria came to Mauritius to the process whereby malaria subsequently became endemic on the island. He noted "its multitudinous small marshes, its grass-ridden streams, and stagnant pools which form on either side of the river-beds after the heavy rains are over". It was in this aquatic

[12] Anderson here cites his 1890 lecture on these events, to which I refer below.

[13] I have found no other reference to this ship, to which Anderson attributes so much significance. Boodhoo (2010, 185) mentions an unnamed "fever stricken ship that had been quarantined off the coast of Albion four months earlier" than the November 1865 outbreak of malaria at Albion estate, located a short way south of Port Louis near Petite Rivière. Perhaps, this ship was the iconic *Spunky*.

environment that the *Anopheles* mosquito flourished. Furthermore, he associated continued Indian immigration and trade with Madagascar, two regions of the Indian Ocean world with long histories of endemic malaria, with maintaining a ready supply of malarial infection (ibid., 178).[14]

To my way of thinking, Anderson's one-to-one conclusion begs the question by taking the Anopheline mosquito as an unproblematic environmental given just waiting to be infected by the Indian carriers of the malaria parasite. But if, as in the case of chikungunya—and as I have just suggested—this insular mosquito population had already undergone certain genetic micro-transformations that rendered them more receptive to the *Plasmodium falciparum* parasite carried in the blood of Indian immigrants from Calcutta, regardless of whether or not they were among the *Spunky*'s passengers, then the sudden eruption of this devastating epidemic makes perfect sense and significantly complicates received interpretations. In addition, since the *Spunky* reached Mauritius in 1865, why did it take a full year for the epidemic to take hold on the island?

An important corollary of the 1866–1867 malaria epidemic in Mauritius was its spread to La Réunion in 1868–1869. Like Mauritius, this French island colony had experienced earlier varieties of vaguely identified relapsing fever, as well as other epidemic diseases such as smallpox and cholera, but never anything identified as malaria. Similar to the situation on its sister island, it initially spread along the low coastal plain, then into the foothills of the island and within three years had become endemic. Not surprisingly, once the mosquito vector had been identified, the main focus of malaria research concentrated on eradication and entered the larger realm of public health policy (see Parahoo 1986; Tchen et al. 2006). More recently, French and Mauritian researchers have sought to understand the specific mechanisms of how malaria came to the Mascarenes. They recognized that they first needed to understand how the mosquito vector established itself on the islands, which they hypothesized occurred in two stages: first the arrival, and second the indigenization of the *Anopheles* (Julvez, Mouchet, Ragavoodoo 1990, 254). Because of the length of time it took to reach the Mascarenes by sail, the short life span (ranging from 5 to 14 days for all four life stages) of the female *Anopheles* mosquitoes and the need for a stable reservoir of fresh water for their eggs to develop, the

[14] This link between the study of disease history and labour migration in Mauritius is a topic that Yoshina Hurgobin (2016) has discussed in a pioneering collection of essays on medicine in the Indian Ocean World (Winterbottom and Tesfaye eds. 2016).

authors ruled out sailing ships—even from Madagascar, where malaria had long been endemic—as a viable means of transportation. What enabled migration across the open waters of the southwestern Indian Ocean and thus facilitated the transportation of the mosquito vectors was, the Franco-Mauritian group deduced, the opening of a regular steamship service by Messageries Maritimes in 1864 that linked the islands to both Madagascar and Africa, without the delays involved in sailing (Julvez et al. 1990, 255; Julvez 1995, 357).[15] For Réunion, Julvez and Mouchet propose in addition that "the theory of wind transport during the trade wind season, since Réunion is located 'leeward' of Mauritius…is supported by the fact that the malaria epidemic started in the North-West of the island facing Mauritius and not near the anchorages" (Julvez and Mouchet 1996, 164).

As research on mosquitoes advanced, investigators sought to identify more accurately than previously the specific variety of *Anopheles gambiae*, which is today the principal vector for malignant malaria globally, that was the original vector. But the existence of four freshwater and two saltwater species of this mosquito, combined with the impact of the mosquito eradication programme on both islands, has complicated matters. Thus, their findings proved elusive except to conclude that "the exclusively African origin of the vectors cannot be doubted" (Julvez et al. 1990, 255). Perhaps of equal interest is their argument that human environmental transformation was an essential factor in preparing the islands for the implantation of malaria. In the case of the two major Mascarene islands (but not Rodrigues or the Seychelles), "the implantation of *Anopheles gambiae s.l.* was consolidated by the deforestation that followed the development of the monoculture of sugar cane, culminating towards 1860" (ibid., 256). In a word, the agricultural transformation of Mauritius for sugarcane plantations, which included road and railway construction, as well as field clearance and the digging of irrigation channels, created an extensive watery environment that was more conducive to the reproduction of the *Anopheles* mosquito (see Julvez et al. 1990, 256; Julvez and Mouchet 1996; Julvez et al. 1998; Boodhoo 2010, 53–54).[16] This is an important general point made by Randall Packard (2007, 12–15 and passim) in his global history of malaria and by Benjamin Reilly (2015) in his case study of malaria in Arabia, as well

[15] For a parallel case of faster steamship travel being the source for the introduction of malignant malaria to northeast Brazil, see Oliveira-Ferreira et al. (2010).

[16] For the Seychelles, see Robert et al. (2011). For the parallel extinction of indigenous birds on the Mascarenes as a consequence of "anthropogenic activities", see Hume (2013).

as by Klein and Samanta for Bengal. Nevertheless, these findings still do not answer unambiguously either how the *Anopheles* mosquito reached the Mascarene islands or, contrarily, whether and how endemic mosquito populations mutated to become vectors for the malarial parasite. For example, in a study of genetic differentiation among geographically distant populations of mosquito vectors, one team of researchers argues that mosquito migration is quite possible across great distances, including oceans (Simard et al. 1999, 1006). In a word, steamships were not necessary for the introduction of *Anopheles* mosquitoes to the Mascarenes. To push these ideas further, closer inspection of the incomplete pieces of information regarding how malaria became an endemic crowd disease in Mauritius in particular raises unanswered questions about chronologies, disease descriptions and the assumptions underlying received interpretations.

Other Possibilities?

There are numerous additional instances of how various epidemic diseases spread around the Indian Ocean, including the Mascarenes. Following up on Myron Echenberg's call for historians to pay more attention to Christie's study of cholera (see footnote 1), a more in-depth exploration of the impact of that disease on the Mascarenes might be essayed. Anderson (1918, 111–151) includes a long presentation on cholera in Mauritius, while medical historians B.A. Gaüzère and P. Aubry (2012) have written about its course on Réunion in the nineteenth century. Although the chronologies of some of these outbreaks in the Mascarenes differ from that described by Christie for East Africa, in other cases there is a direct link, as in the introduction of cholera to Réunion by a boatload of so-called *engagés* (indentured labourers) arriving from Kilwa in 1859 (Echenberg 2011, 55–56; Christie 1876, 113–116; Role 1974).

Another intriguing example is a disease called "Le Barbiers" that repeatedly struck Réunion and other regions of the Indian Ocean world in the first half of the nineteenth century. According to Gaüzère and Aubry, neither the name nor the origins of the disease are known. They note further that it is more than possible that this disease may describe a number of different syndromes (Gaüzère and Aubry 2014, quoted from Abstract). Here again we can see the problem posed by the confusion between dengue and chikungunya: as historians we depend critically on eyewitness and contemporary reported descriptions of whatever it is that we study, in this case epidemic disease, yet words alone are not sufficient for historical analysis.

The literature on chikungunya and malaria suggests to me that, even when we have the full apparatus of modern medical and biogenetic science at our disposal, the identification of epidemic diseases and their dispersion is not always easy. Locating a historically grounded pairing of a disease, whether a virus or a parasite, and its vector(s) is equally challenging. Circumstances matter greatly. In the cases of chikungunya and malaria, even when the terrain is familiar, the mosquitoes that serve as disease vectors are mutable. Moreover, where viruses and parasites are concerned, they too are regularly and constantly adapting to new environmental circumstances, as we can observe with the emergence of resistance to antimalarial drugs, especially by the most lethal form of the malaria parasite, *P. falciparum* (White 2004; Cui et al. 2015). For me, at least, it is this astounding adaptability of both disease and vector that emerges as the most challenging aspect of reconstructing the medical history of disease dispersion in the Indian Ocean World.

Acknowledgements I would like to thank Cal Margulis for his usual efficient research assistance and Ruby Bell-Gam, Librarian for African Studies and International Development Studies, for always tracking down difficult to access material for me through the Young Research Library at UCLA. I am also grateful to my brother, David H. Alpers, M.D., Emeritus William B. Kountz Professor of Medicine, Washington University School of Medicine, for his critical reading of an earlier draft of this chapter and for directing me to some important new sources. Finally, I am indebted to the two organizers of this conference, Gwyn Campbell and Burkhard Schnepel, for their incisive comments on and inspiring suggestions for this chapter.

References

Anderson, Daniel E. 1918. *The Epidemics of Mauritius with a Descriptive and Historical Account of the Island.* London: H.K. Lewis.

Anon. 1892. James Christie Obituary. *The British Medical Journal*, January 9: 97. http://www.ncbi.nlm.nih.gov/pmc/articles/PMC2419488/pdf/brmedj08829-0045.pdf. Accessed 28 July 2016.

———. 2007. Chikungunya in La Reunion – New Findings. *News Medical*, December 10. http://www.news-medical.net/news/2007/12/10/33310.aspx. Accessed 28 July 2016.

Arias-Goeta, Camilo, Laurence Mousson, François Rougeon, and Anna-Bella Failloux. 2013. Dissemination and Transmission of the E1-226V Variant of Chikungunya Virus in *Aedes albopictus* Are Controlled at the Midgut Barrier Level. *PLOS One* 8 (2): e57548. https://doi.org/10.1371/journal.pone.0057548.

Arnold, David. 1991. The Indian Ocean as a Disease Zone, 1500–1950. *South Asia: Journal of South Asian Studies* 14 (2): 1–21.
———. 1999. An Ancient Race Outworn: Malaria and Race in Colonial India, 1830–1930. In *Race, Science and Medicine, 1700–1900*, ed. Waltraud Ernst and Bernard Harris, 123–143. London: Routledge.
Boodhoo, Raj. 2010. *Health, Disease and Indian Immigrants in Nineteenth Century Mauritius*. Port Louis: Aapravasi Ghat Trust Fund.
British Library, Endangered Archives Programme: Project Overview. 2016. EAP863: Preserving a Unique Archive of Diaspora and Disease in the Indian Ocean from 1868 to 1930: A Test Case from Mauritius. http://eap.bl.uk/database/overview_projecta4d?projID+EAP863;r+140. Accessed 29 June 2016.
Burt, Felicity J., Michael S. Ralph, Nestor E. Rulli, Suresh Mahalingam, and Mark T. Heise. 2012. Chikungunya: A Re-emerging Virus. *The Lancet* 379 (9816): 662–671.
Cao-Lormeau, Van-Mai. 2016. Tropical Islands as New Hubs for Emerging Arboviruses. *Emerging Infectious Diseases* 22 (5): 913–915. https://doi.org/10.3201/eid2205.150547.
Carey, Donald E. 1971. Chikungunya and Dengue: A Case of Mistaken Identity? *Journal of the History of Medicine and Allied Sciences* 26 (3): 243–262.
Cherian, Sarah S., Atul M. Walimbe, Santosh M. Jadhav, Swati S. Gandhe, Supriya L. Hundekar, Akhilesh C. Mishra, and Vidya A. Arankalle. 2009. Evolutionary Rates and Timescale Comparison of Chikungunya Viruses Inferred from a Whole Genome/E1 Gene with Special Reference to the 2005–07 Outbreak in the Indian Subcontinent. *Infection, Genetics and Evolution* 9: 16–23.
Christie, James. 1871. Notes on the Cholera Epidemics in East Africa. *The Lancet* 1: 113–115, 186–188.
———. 1872. Remarks on "Kidinga Pepo", a Peculiar Form of Exanthematous Disease, Epidemic in Zanzibar, East Coast of Africa, from July 1870 till January 1871. *Transactions of the Medical and Physical Society of Bombay* 11 (New Series, for the year 1871): 21–31.
———. 1876. *Cholera Epidemics in East Africa*. London: Macmillan.
———. 1881. On Epidemics of Dengue Fever: Their Diffusion and Etiology. *The Glasgow Medical Journal* 3: 161–176.
Cui, Liwang, Sungano Mharakurwa, Daouda Ndiaye, Pradipsinh K. Rathod, and Philip J. Rosenthal. 2015. Antimalarial Drug Honour Resistance: Literature Review and Activities and Findings of the ICEMR Network. *American Journal of Tropical Medicine and Hygiene* 93 (Supplement 3): 57–68.
Deerpalsingh, S., and Marina Carter. 1996. *Select Documents on Indian Immigration: Mauritius, 1834–1926*. Vol. 2. Moka: Mahatma Gandhi Institute.
Delatte, H., L. Bagny, C. Brengue, A. Bouetard, C. Paupy, and D. Fontenille. 2011. The Invaders: Phylogeography of Dengue and Chikungunya Viruses Aedes Vectors, on the South West Islands of the Indian Ocean. *Infection, Genetics and Evolution* 11 (7): 1769–1781.

Echenberg, Myron. 2011. *Africa in the Time of Cholera: A History of Pandemics from 1817 to the Present*. Cambridge: Cambridge University Press.

Ève, Prosper. 2009. Image de l'île par les médecins de marine au début du XIXe siècle. In *Idées et représentations coloniales dans l'Océan Indien*, ed. Norbert Dodille, 509–534. Paris: PUPS.

Floate, H.F.G. 1969. The Mauritian Malaria Epidemic 1866–1868. Geographical Determinism One Hundred Years Ago. *The Journal of Tropical Geography* 29: 10–20.

Gaüzère, B.A., and P. Aubry. 2012. Les épidémies de choléra à l'île de La Réunion au XIXe siècle. *Médecine et Santé Tropicales* 22 (2): 131–136.

———. 2014. La maladie appellée 'Le Barbiers' au XIXe siècle. *Médecine et Santé Tropicales* 24 (3): 241–246.

Gaüzère, B.A., P. Gérardin, D. Vandroux, and P. Aubry. 2012. L'infection à virus chikungunya dans l'Océan Indien: leçons et perspectives. *Médicine Tropicale: revue du Corps santé colonial* 72 (Special Number): 6–12.

Gérardin, Patrick, Vanina Guernier, Joëlle Perrau, Adrian Fianu, Karin Le Roux, Philippe Grivard, Alain Michault, Xavier de Lamballerie, Antoine Flahualt, and François Favier. 2008. Estimating Chikungunya Prevalence in La Réunion Island Outbreak by Serosurveys: Two Methods for Two Critical Times of the Epidemic. *BMC Infectious Diseases* 8: 99. https://doi.org/10.1186/1471-2334-8-99.

Gudo, Eduardo Samo, Gabriela Pinto, Sirkka Vene, Arcildo Mandlaze, Argentina Felisbela Muianga, Julie Cliff, and Kerstin Falk. 2015. Serological Evidence of Chikungunya Virus Among Acute Febrile Patients in Southern Mozambique. *PLOS Neglected Tropical Diseases* 9 (10). https://doi.org/10.1371/journal.pntd.0004146.

Hartwig, Gerald W. 1975. Economic Consequences of Long-Distance Trade in East Africa: The Disease Factor. *African Studies Review* 18 (2): 63–73.

Her, Zhisheng, Yiu-Wing Kam, Raymond T.P. Lin, and Lisa F.P. Ng. 2009. Chikungunya: A Bending Reality. *Microbes and Infection* 11: 1165–1176.

Hume, Julian P. 2013. A Synopsis of the Pre-Human Avifauna of the Mascarene Islands. In *SAPE Proceedings 2013*, ed. Ursula B. Gölich and Andreas Kroh, 195–237. Vienna: Naturhistorisches Museum.

Hurgobin, Yoshina. 2016. Making Medical Ideologies: Indentured Labor in Mauritius. In *Histories of Medicine and Healing in the Indian Ocean World, Volume Two: The Modern Period*, ed. Anna Winterbottom and Facil Tesfaye, 1–26. Basingstoke: Palgrave Macmillan.

Issa, Amina A. 2006. Dhows and Epidemics in the Indian Ocean Ports. *Ziff Journal* 4: 63–70.

Jansen, Karine Aasgaard. 2016. Tropical Disease and the Making of France in Réunion. In *Histories of Medicine and Healing in the Indian Ocean World, Volume Two: The Modern Period*, ed. Anna Winterbottom and Facil Tesfaye, 147–169. Basingstoke: Palgrave Macmillan.

Josseran, Loïc, Christophe Paquet, Abdelkrim Zehgnoun, Nadège Caillere, Alain Le Tertre, Jean-Louis Solet, and Martine Ledrans. 2006. Chikungunya Disease Outbreak, Reunion Island. *Emerging Infectious Diseases* 12 (12): 1994.

Julvez, Jean. 1995. Historique du paludisme insulaire dans l'océan Indien (partie sud-ouest): Une approache éco-épiémiologique. *Cahiers Santé* 5 (6): 353–357.

Julvez, Jean, and J. Mouchet. 1996. Anophelism and Epidemiological Patterns of Malaria in the South-West Indian Ocean Archipelagos. *Research and Reviews in Parasitology* 56 (4): 161–167.

Julvez, Jean, J. Mouchet, and C. Ragavoodoo. 1990. Epidemiologie historique du paludisme dans l'archipel des mascareignes (Océan Indien). *Annales des Sociétés belges de Médicine tropicale, de parasitologie et de mycologie humaine et animale* 70 (4): 249–261.

Julvez, Jean, C. Ragavoodoo, A.R. Gopaul, and J. Mouchet. 1998. Maladies humaines transmises par les culicidés dans les Îles du sud-ouest de l'Océan Indien. *Bulletin de la Société de Pathologie Exotique* 91 (1): 99–103.

Kajeguka, Debora C., Robert D. Kaaya, Steven Mwakalinga, Rogathe Ndossi, Arnold Ndaro, Jaffu O. Chilongola, Franklin W. Mosha, Karin L. Schiøler, Reginald A. Kavishe, and Michael Alifrangis. 2016. Prevalence of Dengue and Chikungunya Virus Infections in North-Eastern Tanzania: A Cross Sectional Study Among Participants Presenting with Malaria-like Symptoms. *BMC Infectious Diseases* 16: 183. https://doi.org/10.1186/s12879-016-1511-5.

Kampango, Ayubo, and Ana Paula Abílio. 2016. The Asian Tiger Hunts in Maputo City – The First Confirmed Report of *Aedes (Stegomyia) albopictus* (Skuse, 1895) in Mozambique. *Parasites & Vectors* 9 (76). https://doi.org/10.1186/s13071-016-1361-4.

Karamanou, Marianna, George Panayiotakopoulos, Gregory Tsoucalas, Antonis A. Kousoulis, and George Androutsos. 2012. From Miasmas to Germs: A Historical Approach to Theories of Infectious Disease Transmission. *Le infezioni in Medicina: rivista periodica di eziologia, epidemiologia, diagnostica, clinica e terapia delle patologie infettive* 20 (1): 58–62. https://www.researchgate.net/publication/223957556_From_miasmas_to_germs_A_historical_approach_to_theories_of_infectious_disease_transmission. Accessed 14 Feb 2017.

Khan, K., I. Bogoch, J.S. Brownstein, J. Miniota, A. Nicolucci, W. Hu, E.O. Nsoesie, et al. 2014. Assessing the Origin of and Potential for International Spread of Chikungunya Virus from the Caribbean. *PLOS Currents Outbreaks*, June 6, Edition 1. https://doi.org/10.1371/currents.outbreaks.2134a0a7bf37fd8d388181539fea2da5.

Kjekshus, Helge. 1996 [1977]. *Ecology Control and Economic Development in East African History: The Case of Tanganyika 1850–1950*. London: James Currey.

Klein, Ira. 2001. Development and Death: Reinterpreting Malaria, Economics and Ecology in British India. *The Indian Economic and Social History Review* 38 (2): 147–179.

Knoll, Eva-Maria. 2018. "Considering the Island Capital Male" as a Hub for Health-Related Mobilities'. In *Connectivity in Motion: Island Hubs in the Indian Ocean World*, ed. Burkhard Schnepel and Edward A. Alpers, 319–343. London: Palgrave Macmillan.

Koponen, Juhani. 1988. *People and Production in Late Precolonial Tanzania: History and Structures*. Helsinki: Finnish Society for Development Studies/Finnish Historical Society/Scandinavian Institute of African Studies.

Kumar, Abhishek, Prabhudutta Mamidi, Indrani Das, Tapas K. Nayak, Sameer Kumar, Jagamohan Chhatai, Subhasis Chattopadhyay, Amol R. Suryawanshi, and Soma Chattopadhyay. 2014. A Novel 2006 Indian Outbreak Strain of Chikungunya Virus Exhibits Different Pattern of Infection as Compared to Prototype Strain. *PLOS One* 9 (1): e85714. https://doi.org/10.1371/journal.pone.0085714.

Kuno, Goro. 2015. A Re-Examination of the History of Etiologic Confusion Between Dengue and Chikungunya. *PLOS Neglected Tropical Diseases* 9 (11): e0004101. https://doi.org/10.1371/journal.pntd.0004101.

Lahariya, Chandrakant, and S.K. Pradhan. 2006. Emergence of Chikungunya Virus in Indian Subcontinent After 32 Years: A Review. *Journal of Vector Borne Diseases* 43: 151–160.

Lam, S.K., K.B. Chua, P.S. Hooi, M.A. Rahimah, S. Kumari, M. Tharmaratnam, S.K. Chuah, D.W. Smith, and I.A. Sampson. 2001. Chikungunya Infection – An Emerging Disease in Malaysia. *Southeast Asian Journal of Tropical Medicine and Public Health* 32 (3): 447–451.

Laras, Kanti, Nono C. Sukri, Ria P. Larasati, Michael J. Bangs, Rizal Kosim, Tony Wandra Djauzi, John Master, et al. 2005. Tracking the Re-Emergence of Epidemic Chikungunya Virus in Indonesia. *Transactions of the Royal Society of Tropical Medicine and Hygiene* 99 (2): 128–141.

Morrison, Clayton, R. Kenneth, S. Plante, and Mark T. Heise. 2016. Chikungunya Virus: Current Perspectives on a Reemerging Virus. *Microbiology Spectrum* 4 (3): EI10-0017-2016: 3. https://doi.org/10.1128/microbiolspec.E|10-0017-2016.

Mukherjee, Siddhartha. 2016. The Chase: How Fast Can We Roll Out a Zika Vaccine? *The New Yorker*, August 22, 32–37.

Mukherjee, Sujata. 2008. Environmental Thoughts and Malaria in Colonial Bengal: A Study in Social Response. *Economic and Political Weekly* 43 (12/13): 54–61.

Ng, Lee Ching, and C. Hapuarachchige. 2010. Tracing the Path of Chikungunya Virus – Evolution and Adaptation. *Infection, Genetics and Evolution* 10: 876–885.

Njenga, M. Kariuki, L. Nderitu, J.P. Ledermann, A. Ndirangu, C.H. Logue, C.H.L. Kelly, R. Sang, K. Sergon, R. Breiman, and A.M. Powers. 2008. Tracking Epidemic Chikungunya Virus into the Indian Ocean from East Africa. *Journal of General Virology* 89 (11): 2754–2760. https://doi.org/10.1099/vir.0.2008/005413-0.

Oliveira-Ferreira, J., M.V. Lacerda, P. Brasil, J.L. Ladislau, P.L. Tauil, and C.T. Daniel-Ribeiro. 2010. Malaria in Brazil: An Overview. *Malaria Journal* 9: 115. https://doi.org/10.1186/1475-2875-9-115.

Packard, Randall M. 2007. *The Making of a Tropical Disease: A Short History of Malaria*. Baltimore, MD: The Johns Hopkins University Press.

Paganin, Fabrice, Giannandrea Borgherini, Frédéric Staikowsky, and Claude Arvin-Berod. 2006. Chikungunya à l'île de la Réunion: chroniqiue d'une épidémie annoncée. *Presse Médicale* 35: 641–646.

Parahoo, Kader A. 1986. Early Colonial Health Developments in Mauritius. *International Journal of Health Services* 16 (3): 409–423.

Paupy, C., H. Delatte, L. Bagny, V. Corbel, and D. Fontenille. 2009. *Aedes albopictus*, An Arbovirus Vector: From the Darkness to the Light. *Microbes and Infection* 11 (14–15): 1177–1185. https://doi.org/10.1016/j.micinf.2009.05.005.

Pialoux, Gilles, Bernard-Alex Gaüzère, and Michel Strobel. 2006. Infection à virus Chikungunya: revue génerale par temps d'épidémie [Chikungunya Virus Infection: Review Through an Epidemic]. *Médecine et maladies infectieuses* 36: 253–263. https://doi.org/10.1016/j.medmal.2006.04.002.

Pialoux, Gilles, Bernard-Alex Gaüzère, Stéphane Jauréguiberry, and Michel Strobel. 2007. Chikungunya, an Epidemic Arbovirosis. *Lancet Infectious Disease* 7: 319–327.

Pierre, Vincent, Laurent Filleul, Jean-Louis Solet, Philippe Renault, Daouda Sissoko, and Christian Lassalle. 2006. Épidémie de Chikungunya à l'île de la Réunion, 2005–2006. *Presse Médicale* 35: 1188–1189.

Reilly, Benjamin. 2015. *Slavery, Agriculture, and Malaria in the Arabian Peninsula*. Athens: Ohio University Press.

Renault, Philippe, Jean-Louis Solet, Daouda Sissoko, Else Balleydier, Sophie Larrieu, Laurent Filleul, Christian Lassalle, et al. 2007. A Major Epidemic of Chikungunya Virus Infection on Réunion Island, France, 2005–2006. *American Journal of Tropical Medicine and Hygiene* 77 (4): 727–731.

Renault, P., E. Balleydier, E. D'Ortenzio, M. Bâville, and L. Filleul. 2012. Epidemiology of Chikungunya Infection on Reunion Island, Mayotte, and Neighboring Countries. *Médecine et Maladies Infectieuses* 42: 93–101. https://doi.org/10.1016/j.medmal.2011.12.002.

Robert, Vincent, Gérard Rocamora, Simon Julienne, and Steven M. Goodman. 2011. Why Are Anopheline Mosquitoes Not Present in the Seychelles? *Malaria Journal* 11 (10): 31. https://doi.org/10.1186/1475-2875-10-31.

Robinson, Marion C. 1955. An Epidemic of Virus Disease in Southern Province, Tanganyika Territory in 1952–53. Clinical Features. *Transactions of the Royal Society of Tropical Medicine and Hygiene* 49 (1): 28–32.

Rockel, Stephen J. 2006. *Carriers of Culture: Labor on the Road in Nineteenth-Century East Africa*. Portsmouth, NH: Heinemann.

Role, A. 1974. L'Affaire' du 'Mascareignes': le choléra à La Réunion en 1859. *Histoire des Sciences Médicales* 8: 503–526. http://192.168.1.1:8181/http://www.biusante.parisdescartes.fr/sfhm/hsm/HSMx1974x008x003/HSMx1974x008x003x0503.pdf. Accessed 9 Aug 2016.

Ross, Ronald. 1908. *Report on the Prevention of Malaria in Mauritius*. London: Waterlow and Sons.

Ross, R.W. 1956. The Newala Epidemic, III: The Virus; Isolation, Pathogenic Properties and Relationship to the Epidemic. *Journal of Hygiene* 54 (2): 177–191. http://192.168.1.1:8181/http://www.biusante.parisdescartes.fr/sfhm/hsm/HSMx1974x008x003/HSMx1974x008x003x0503.pdf.

Samanta, Arabinda. 2002. *Malarial Fever in Colonial Bengal, 1820–1939: Social History of an Epidemic*. Kolkata: Firma KLM.

Schuffenecker, Isabelle, Isabelle Iteman, Alain Michault, Séverine Murri, Lionel Frangeul, Marie-Christine Vaney, Rachel Lavenir, et al. 2006. Genome Microevolution of Chikungunya Viruses Causing the Indian Ocean Outbreak. *PLOS Medicine* 3 (7): e263., 1058–1059, 1067–1068. https://doi.org/10.1371/journal.pmed.0030263.

Sergon, Kibet, Ali Ahmed Yahaya, Jennifer Brown, Said A. Beja, Mohammed Mlindasse, Naphtali Agata, Yokouide Allaranger, et al. 2007. Seroprevalence of Chikungunya Virus Infection on Grande Comore Island, Union of the Comoros. *American Journal of Tropical Medicine and Hygiene* 76 (6): 1189–1193.

Sergon, Kibet, Charles Njuguna, Rosalia Kalani, Victor Ofula, Clayton Onyango, Limbaso S. Konongoi, Sheryl Bedno, et al. 2008. Seroprevalence of Chikungunya Virus (CHIKV) Infection on Lamu Island, Kenya, October 2004. *American Journal of Tropical Medicine and Hygiene* 78 (2): 333–337.

Simard, Frederic, Didier Fontenille, Tovi Lehmann, Romain Girod, Laurent Brutus, Raheem Gopaul, Christian Dournon, and Frank H. Collins. 1999. High Amounts of Genetic Differentiation Between Populations of the Malaria Vector *Anopheles arabiensis* from West Africa and Eastern Outer Islands. *American Journal of Tropical Medicine and Hygiene* 60 (6): 1000–1009.

Simon, F., H. Tolou, and P. Jeandel. 2006. Chikungunya, l'épidémie que l'on n'attendait pas/The Unexpected Chikungunya Outbreak. *La Revue de médecine interne* 27: 437–441. https://doi.org/10.1016/j.revmed.2006.03.028.

Tchen, J., A. Ouledi, J.F. Lepère, D. Ferrandiz, and J.L. Yvin. 2006. Epidemiology and Prevention of Malaria in the Southwestern Islands of the Indian Ocean. *PubMed – Med Trop (Mars)* 66 (3): 295–301. http://www.ncbi.nlm.nih.gov/pubmed/?term=tchen+j%2C+malaria. Accessed 22 Aug 2016.

Tsetsarkin, Konstantin A., and Scott C. Weaver. 2011. Sequential Adaptive Mutations Enhance Efficient Vector Switching by Chikungunya Virus and Its Epidemic Emergence. *PLOS Pathogens* 7 (12): e1002412. https://doi.org/10.1371/journal.ppat.1002412.

Tsetsarkin, Konstantin A., Dana L. Vanandingham, Charles E. McGee, and Stephen Higgs. 2007. A Single Mutation in Chikungunya Virus Affects Vector Specificity and Epidemic Potential. *PLOS Pathogens* 3 (12): e201. https://doi.org/10.1371/journal.ppat.0030201.

Tsetsarkin, Konstantin A., Rubing Chen, Ruimei Yun, Shannon L. Rossi, Kenneth S. Plante, Mathilde Guerbois, Naomi Forrester, et al. 2014. Multi-peaked Adaptive Landscape for Chikungunya Virus Evolution Predicts Continued Fitness Optimization in *Aedes albopictus* Mosquitoes. *Nature Communications* 5 (4084). https://doi.org/10.1038/ncomms5084, https://www.nature.com/articles/ncomms5084.

Tsetsarkin, Konstantin A., Rubing Chen, and Scott C. Weaver. 2016. Interspecies Transmission and Chikungunya Virus Emergence. *Current Opinion in Virology* 16: 143–150. https://doi.org/10.1016/j.coviro.2016.02.007.

Vairo, Francesco, Emanuele Nicastri, Silvia Meschi, Monica Sanè Schepisi, Maris Grazia Paglia, Nazario Bevilacqua, Sabina Mangi, et al. 2012. Seroprevalence of Dengue Infection: A Cross-Sectional Survey in Mainland Tanzania and on Pemba Island, Zanzibar. *International Journal of Infectious Diseases* 16: e44–e46. https://doi.org/10.1016/j.ijid.2011.09.018.

Weaver, Scott C., and Naomi L. Forrester. 2015. Chikungunya: Evolutionary History and Recent Epidemic Spread. *Antiviral Research* 120: 32–39. https://doi.org/10.1016/j.antiviral.2015.04.016.

Weaver, Scott C., and Marc Lecuit. 2015. Chikungunya and the Global Spread of a Mosquito-Borne Disease. *The New England Journal of Medicine* 372 (13): 1231–1239.

White, Nicholas J. 2004. Antimalarial Drug Resistance. *The Journal of Clinical Investigation* 113 (8): 1084–1092.

Winterbottom, Anna, and Facil Tesfaye, eds. 2016. *Histories of Medicine and Healing in the Indian Ocean World, Volume Two: The Modern Period*. Basingstoke: Palgrave Macmillan.

Yergolkar, Prasanna N., Babasaheb V. Tandale, Vidya A. Arankalle, Padmakar S. Sathe, A.B. Sudep, Swati S. Gandhe, Mangesh D. Gokhle, George P. Jacob, Supriya L. Hundekar, and Akhilesh C. Mishra. 2006. Chikungunya Outbreaks Caused by African Genotype, India. *Emerging Infectious Diseases* 12 (10): 1580–1583.

CHAPTER 10

Challenging Chikungunya: Resistance to Public Health Measures and Aetiology During the 2005–2007 Epidemic in Réunion

Karine Aasgaard Jansen

Cyclical monsoon winds have long carried ships across the Indian Ocean. In their turn, vessels have facilitated the diffusion of both human and nonhuman cargoes such as vectors and viruses.[1] Indeed, disease—the focus of this volume—has constituted an integral part of the extensive network of maritime exchange and migration of the Indian Ocean World (IOW). Scholarly studies of the IOW exchange of people and pathogens include David Arnold's seminal article 'The Indian Ocean as a disease zone, 1500–1950' (1991) and the anthology 'Histories of Medicine and Healing in the Indian Ocean World' (Winterbottom and Tesfaye 2016). Travelling disease is, thus, neither a new empirical phenomenon nor only a current interest within IOW studies. This regional disease diffusion was demonstrated in the

[1] Vectors are living organisms that can transmit infectious diseases between humans or from animals to humans. Many of these vectors are bloodsucking insects, such as mosquitoes.

K. A. Jansen (✉)
Umeå University, Umeå, Sweden
e-mail: karine.jansen@umu.se

G. Campbell, E.-M. Knoll (eds.), *Disease Dispersion and Impact in the Indian Ocean World*, Palgrave Series in Indian Ocean World Studies, https://doi.org/10.1007/978-3-030-36264-5_10

2004–2007 epidemic of the vector-borne disease of chikungunya, the first ever to hit the IOW. The epidemic spread from Kenya's Swahili coast to the Western Indian Ocean islands of Lamu, the Seychelles, Madagascar, the Comoros, Mayotte, Réunion, and Mauritius, followed by India. In addition, imported cases from exposed travellers returning from the affected areas were later identified in France, Italy, Hong Kong, USA, and Canada (Alpers in this volume; Taglioni and Dehecq 2009; Njenga et al. 2008).

In this study, which is based on a total of eight months of ethnographic fieldwork in the island of Réunion from 2009 to 2010, I will discuss how human-environment interaction in a traditional Réunionese household (*kaz creole*) and its adjacent garden (*jardin creole*) played a central role in forming local perceptions of chikungunya and resistance to public health interventions—and aetiology.[2] My main argument is that many Réunionese interpreted chikungunya, a new disease to the island, within preexisting narrative frameworks of vector-borne disease diffusion and prevention that had developed in response to the malaria outbreaks of the 1950s. I will investigate how the *Aedes* mosquito's breeding grounds in a *jardin creole*, as well as previous public health interventions against vector-borne diseases, contributed to the stigmatisation of chikungunya. Moreover, despite widespread and easy access to public health information regarding chikungunya's mode of transmission, I will argue that stigmatisation may have led many Réunionese to reject the idea that chikungunya was a vector-borne disease.

Chikungunya: An Outline of the Epidemic in Réunion

Chikungunya was first identified in 1953 when the virus was isolated from the blood of a febrile Makonde speaker on the border between Tanzania and Mozambique (Pialoux et al. 2007, 319; see also Alpers in this volume). Chikungunya attacks the joints, leading to painful swelling and considerably reduced motor function that may last for a few days up to several years. The virus is carried by the female *Aedes aegypti* and *albopictus* mosquito, which is also responsible for spreading dengue, yellow fever,

[2] In Réunion, *creole* refers to anyone or anything of local Réunionese origin. This is in contrast to, for example, Mauritius where *Creole* rather functions as an ethnic identity marker for Mauritians who are primarily descendants of African and Malagasy slaves (Eriksen 1998).

and zika. As of now, there is no effective vaccine against the virus, but infection results in lifelong immunity.

According to the French Institute for Public Health Surveillance (*Institut de veille sanitaire*, INVS), the chikungunya epidemic began in Kenya in June 2004 and reached Réunion in March 2005. One month later, its presence was communicated to the general public by the Regional Department of Health and Social Affairs (*Direction régionale des affaires sanitaires et sociales* (DRASS); Watin 2008, 2010; Weinstein and Ravi 2009), the governmental agency responsible for vector control and eradication.[3] DRASS assured the Réunionese population that chikungunya was nonfatal and that the epidemic, which authorities were closely monitoring, would be over by the end of July 2005 (Watin 2008, 243, 2010; Weinstein and Ravi 2009). None of these statements proved correct. The epidemic peaked in the first two months of 2006, with over 25,000 new cases being registered in the last week of January and an astounding 45,000 in the first week of February (Taglioni and Dehecq 2009, 15). In several cases, severe health complications were reported, and in January 2006 chikungunya claimed its first fatal victim worldwide—a ten-year-old boy (Leyral 2006, 14).

At the peak of the chikungunya epidemic, many Réunionese claimed that the air had been polluted by a medical or military experiment, a chemical outlet, or even through biological terrorism inflicted by Bin Laden and his alleged Comoro Muslim accomplices (Jansen 2016, 163). The reaction from France was different. Following the boy's death, the French government acknowledged the potential gravity of the epidemic and responded with economic, medical, and preventive assistance. In total, 91 million Euros were contributed towards research, sanitary improvements, and financial relief. The government also mobilised 720 locally stationed metropolitan—that is, French mainland—military personnel to conduct so-called demosquitofications (P. D. 2006, 6; Payrard 2006, 14), the local term for prophylactic measures such as the spraying of insecticides. These measures occurred, however, fully a year after the virus's presence had been documented in Réunion and almost two years

[3] In April 2010, DRASS underwent an organisational transition and is now known as the *Agence de santé de l'Océan Indien* (The Health Agency of the Indian Ocean, ARS). ARS is non-governmental, whereas DRASS responded directly to the French prefect. The prefect represents the French national government at a local level. As DRASS was still in operation during both the chikungunya epidemic and my main fieldwork in 2009, I will still refer to DRASS and not ARS throughout this chapter.

after the epidemic first broke out in Kenya. Intentional or unintentional misinformation regarding the severity of the epidemic, and late and inadequate official intervention, depending on one's point of view, prompted significant local criticism of the French authorities, and of France's neglect of its overseas citizens (Audifax 2006). Official response to the epidemic also aroused general public mistrust in the public health administration, DRASS included. This politicisation of the epidemic resulted in a deep division, in which the metropolitan French government found itself opposed by all interests in the island, including the Réunionese regional government, local media and civil institutions, and the general population. By April 2007, when Philippe Bas and Hervé Mariton, respectively French ministers of Health and Overseas Departments, declared the epidemic to be over (Payrard 2007), 266,000 local Réunionese, nearly 30 per cent of the island's total estimated population of 802,000 (Rallu 2009), had been infected by chikungunya and 250 had died (Renault et al. 2008).

Methods and Research Background

With some exceptions (e.g., Jansen 2012, 2013; Watin 2010; Weinstein and Ravi 2008, 2009), most studies of the chikungunya epidemic in Réunion have been by epidemiologists and entomologists who used quantitative methodologies (e.g., Pialoux et al. 2007; Renault et al. 2008; Taglioni and Dehecq 2009). However, as a medical anthropologist, I am interested in how disease and illness are experienced, explained, and treated at a local level—and in order to study this, I have employed qualitative methods, notably the gathering of ethnographic data based upon researcher participation, which provides a good indication of every-day social life as it unfolds in particular contexts. A large part of my fieldwork in Réunion was, for example, spent outdoors in the gardens of 16 informants who had suffered from chikungunya during the epidemic. Due to the island's temperate weather, people spend as much time outdoors as indoors. A *jardin creole* functions as a de facto extension of a *kaz creole*, and people's activities and interactions in them played a central part in this study's data collection and analysis. Conversations with informants were often carried out while performing various household and gardening chores, such as doing laundry, watering plants, and picking and shelling beans, or enjoying a cup of coffee on the *varang* (overbuilt terrace) in front of a *kaz creole*. During an intense week of fieldwork that followed the discovery of three cases of chikungunya in Réunion in August 2009, I also joined a team of DRASS field agents to conduct daily garden inspections for mosquito breeding grounds. This enabled me to learn more about vector-borne diseases in general and chikungunya in particular, DRASS's preventive methods, and how DRASS and the public health policies they implemented were received by the public. From the peak of the epidemic in January 2006, DRASS officials regularly conducted garden inspections as part of their preventive public health measures, and these were still being conducted during my last period of fieldwork in late 2010.

A crucial part of fieldwork is the writing of field notes in order to gather ethnographic knowledge (Emerson et al. 1995). Observations and conversations are meticulously jotted down from notes taken on site, or, as in this case, primarily at home in the evening. I also conducted semi-structured interviews with informants, public health and sanitation staff, entomologists, and epidemiologists, in Creole or French according to their preferences. Furthermore, I analysed various written sources on chikungunya, such as local and national newspapers, public health reports, and preventive campaign material. In addition, I reviewed governmental documents and newspaper reports of the massive anti-malaria campaigns that took place on the island during the 1950s at the Departmental Archives in Réunion's capital, St Denis. This review allowed me to compare the public health discourses on chikungunya with those of a previous vector-borne disease, specifically the discourses surrounding their spread and the preventive measures that were adopted in response to them.

Placing Chikungunya: Disease Transmission, Vector Adaption, and Living Conditions

The location for my fieldwork was the town of St Pierre on the west coast of Réunion. St Pierre has a population of approximately 26,000 and was one of the island's urban centres most affected by the epidemic. Indeed, the first case of chikungunya in Réunion occurred in a St Pierre neighbourhood inhabited by several of my informants, close to a public semi-open-air *lavoir*, or wash house. A *lavoir* consists of a long row of large double washbasins set in stone and in this case was covered by a slightly dilapidated and rusted tin roof. The risk of contracting chikungunya increases with exposure to stagnant water, in both artificial and natural reservoirs, that acts as a breeding ground for the disease vector. Examples of artificial reservoirs of stagnant water include wash basins, such as the ones at the *lavoir*, gutters, flower pots, bottles, bottle caps, plastic food containers, and discarded car tyres, while rain ponds and plant axils typically comprise natural reservoirs (Bonn 2006).

As the *Aedes* mosquito thrives in close-knit spaces provided by artificial reservoirs in densely populated and urban environments (ibid.; Nading 2014), chikungunya affected large sectors of the Réunionese population, above all in areas where people live in close proximity to each other. Moreover, according to Raude and Setbon (2009, 690), there is a well-established link between a person's socioeconomic position and his or her risk of contracting chikungunya. In St Pierre, the neighbourhood surrounding the *lavoir* called *quartier lavoir* is known as a *quartier chaud* (hot area)—a term denoting high local unemployment rates and accompanying social problems.[4] In addition to being used for its original purpose of doing laundry, the particular *lavoir* in question also served a social function as a neighbourhood meeting place. This contributed towards the accumulation of waste near the *lavoir* and, in turn, gave the *Aedes* mosquitoes an additional breeding ground. While the women who lived in the nearby houses, the majority of whom were Mahoran, were busy washing clothes and chatting with each other, men sat in small groups on the grassy slope next to the *lavoir*, drinking rum and eating take-away food, playing cards, dominos, or the guitar, and watching the women work. As much as the *lavoir* itself contributed towards the accumulation of stagnant water in ponds, so did some of the men's leftover rubbish such as bottles, bottle caps, and food containers.

That the first case of chikungunya was identified in the *quartier lavoir* was brought up by several informants living in the area's vicinity. Instead of looking for mosquito breeding grounds in their own gardens, many informants looked 'over the fence' to poorer neighbourhoods for possible mosquito breeding grounds, especially in *quartiers* primarily inhabited by Mahoran immigrants. This contributed to the stigmatisation of St Pierre's large population of Mahorans as responsible for spreading the disease. However, although poor sanitary conditions in economically and socially challenged areas may have facilitated the spread of vector-borne diseases (Castro et al. 2010; Nading 2014; Raude and Setbon 2009), the *Aedes* does not differentiate between different sources of stagnant water. The mosquito can equally breed in the flower pots that characterise well-kept

[4] As of 2010, Réunion had the highest level of unemployment in all of the French departments (over 30 per cent) (Roinsard 2010, 21).

gardens of better-off neighbourhoods, and are the distinguishing feature of a traditional *jardin creole*, the pride and joy of many of my Réunionese informants who lived near the *quartier lavoir*.

Staging the Context of the *Kaz* and *Jardin Creole*

While the concept of a garden may intuitively bring to mind the manicured lawns and meticulous flower beds of suburban America (Jenkins 1994), the purpose of a *jardin creole* in Réunion is not merely decoration or leisure. Instead, a *jardin creole* mirrors the spatial layout of a *kaz creole* and acts as a stage for the unfolding of every-day domestic life in both rural and urban, and privileged and impoverished neighbourhoods. A *kaz creole* is composed of two compartments, and each has a distinct purpose in terms of practicality and comfort. These compartments consist of the *kour devant* (the front garden facing the street) and the *kour derriere* (the garden at the back of the house) (Wolff 1991). The *kour devant* functions as the public area of a *kaz creole* and can be compared to Erving Goffman's (1992 [1959]) dramaturgical understanding of a front stage for social behaviour. This is, for example, where the household commonly receives guests on the *varang* for a cup of sweet tea or coffee, usually served among potted plants, and if space allows it, frangipani, mango, or lychee trees, sometimes with orchids growing off their trunks. In a *jardin creole*, plants are typically potted rather than grown in flower beds. The quantity of potted plants varies from a few to several hundred according to the household's available space and income. With the potted plants as stage props, the performance, in Goffmanian terms, thus concerns both the aesthetic staging of the *kour devant* and the social interactions and activities taking place there. Here, it may be useful to compare the function of Réunionese gardens with those of urban Swedish allotments where, Michel Conan (1999, 200) claims, gardeners function as actors who play out the myth of what they imagine to have been a good rural life in 'the olden times'. This re-enactment is done through the performance of various social activities, such as collective gatherings and sharing traditional meals, and it bears a resemblance to the sipping of coffee on the *varang* in the *kour devant* of a *jardin creole*.

However, a *jardin creole* also consists of a *kour derriere*, which can be seen as the very opposite of the staged *kour devant* in both a literal and an allegorical sense. In the *kour derriere*, public performance is replaced by private undertakings that are considered 'behind the scenes' activities, such as cleaning and drying clothes, food preparation, and other mundane household tasks. Moreover, the stage props found in the *kour devant*, such as pineapple flowers, anthuriums, hibiscus plants, or bougainvillea, are replaced in the *kour derriere* with weeds growing among various discarded items, such as car tyres, bicycles, toys, bottles, gardening tools, and furniture or electrical appliances both in and out of use. The *kour derriere* may also house the household's poultry, dog or bird cages, and storage sheds (Wolff 1991). It thus functions as the back stage of a *kaz creole* and is intended for the household's members only. It was therefore with considerable pride that I accepted an invitation to do laundry on a regular basis with one of my neighbours and key informants in her *kour derriere*. Nevertheless, even then, the distinction between the front and back stages was maintained when it was time for our coffee breaks, which always took place on the *varang*.

The actual *kaz* (the house) is situated between the *kour devant* and the *kour derriere*. Moreover, as Éliane Wolff (1991) notes, such living patterns are common not only in the traditional one-family *kaz creole* but also in terraced houses or blocks of flats. Here, the *kour devant* and the *kour derriere* are represented by, for example, terraces, smaller front- and backyards at street level, or even rooms at the front and the back of an apartment. As potted plants do not require any ground or space to speak of, a *jardin creole* is therefore also common in urban areas, including in social-housing estates.

Gardens and Mosquitoes: Touching on Human-Environment Interaction and Multispecies Ethnography

From a cultural perspective, the Réunionese garden is less about gardening than housekeeping and living in general. For decades, anthropologists have been investigating people's relations with their gardens as part of a wider social system (Hasna 2003). Early anthropological studies of gardens have illustrated the importance of these spaces for economic life, in addition to their symbolic importance as representations of society as a whole (ibid.). A classic example of this line of research is Bronislaw Malinowski's (1935) study 'Coral Gardens and their Magic'. From a functionalist perspective, Malinowski documents the interrelations between the Trobriand Islanders' gardening activities, religious practices, and agricultural economy, all of which he claims form a holistic part of the Trobrianders' organisation of social life.

More recent studies such as Conan (1999) consider that on the one hand gardening reproduces societal structures and on the other illustrates people's relationship to nature (ibid., 202). Contextual human-environment interaction—that is, the interrelations between people and their surroundings—entails how people act on and respond to the environment and how such exchanges influence the social organisation of a given society (Hasna 2003). Mahbuba Hasna (ibid.) claims, for example, that gardening connects people to their plants and environment through a systemic process. However, her fieldwork in Inhambane, Mozambique, leads Julie Archaumbault (2016) to argue that gardening is also an aesthetic and often profoundly affective endeavour. According to Archaumbault (ibid., 246), human-plant relations should be understood not merely as a way in which societies reproduce their social bonds, but also a reflection of genuine love for plants.

While clearly stating the relevance of continued anthropocentric analyses, Archaumbault (ibid., 248) also engages with the growing field of multispecies ethnography, which focuses both on humans and other organisms that are linked to human worlds and livelihoods (Kirksey and Helmreich 2010, 545). The core relationships scrutinised are both human-to-human and human-to-nonhuman. The latter can include gardeners' relationship with their plants, as discussed by Archaumbault (2016), or people's relations with animals, insects, fungi, and microbes. Multispecies ethnography focuses on multispecies mingling in zones of contact where the dichotomy between nature and culture is challenged in ways that generate new ecologies and 'becomings' (Kirksey and Helmreich 2010, 546). In this context, in his analysis of how residents of Ciudad Sandino in Nicaragua share their lives with dengue-carrying *Aedes aegypti* mosquitoes, Alex Nading (2014) uses the concept of 'entanglement'. He argues that in studying the diffusion of dengue, it is both difficult but also analytically unproductive to separate humans from nonhumans, such as mosquitoes, because their lives are so intertwined (ibid., 19). Kirksey and Helmreich (2010, 545) take the concept further by suggesting that such entanglements can be perceived as 'symbiopolitical', whereby a 'multitude of organisms' livelihoods shape and are shaped by political, economic, and cultural forces'. As pointed out by Hugh Raffles (2010), people's relationships with mosquitoes are usually less affectionate than, say, Inhambane gardeners' love for their plants (Archaumbault 2016). Not only are mosquitoes a nuisance to most people, but they can also spread life-threatening diseases (Raffles 2010). Nevertheless, as Nading points out, while the language of risk prompts us to think in terms of the body's proximity to mosquitoes and viruses, risk also comes in the form of the body's entanglement with them (2014, 92). In Réunion, both flower pots in the ***kour devant*** and the accumulation of items in the ***kour derriere*** may lead to a flourishing environment for the *Aedes* mosquito. Mosquitoes, sometimes disease carrying ones, are familiar features of Réunionese landscapes. As M. Noel, a 56-year-old informant, noted:

> The chikungunya mosquito, it lives in the environment where we live, and thus we don't really notice it [...] It's not a new mosquito. When I was a child, I lived in St Rose where I often frequented the vanilla fields. And in those fields there was only this type of mosquito, the white and black mosquito. I have always known the white and black mosquito. (Interview, 14 September 2009)

Several other informants also brought up the fact that the *Aedes* mosquito, with its characteristic white- and black-striped body, was a longstanding and well-known part of Réunion's environment. However, the fact that such a familiar part of the island's ecosystem suddenly began spreading a previously unknown disease led many Réunionese to question the aetiological origin of chikungunya. While mosquitoes were commonly presented to me by research participants as an intrinsic part of Réunionese gardens, chikungunya was often depicted as something alien and aerial (Jansen 2013, 180–181). In the following section, I will discuss how this interpretation of chikungunya was based on previous experiences with public health measures against vector-borne diseases, particularly malaria, and the post-colonial and political discourses surrounding these interventions.

Public Health Interventions and Past and Present Stigmatisation of Vector-Borne Diseases

According to DRASS, about 70 per cent of mosquito breeding grounds are located close to people's homes.[5] Consequently, public health interventions have been primarily directed towards the control and eradication

[5] This information is from the DRASS preventive campaign pamphlet '*Adoptons les bons gestes*!'/'Embrace the right actions!'.

of mosquito breeding grounds through so-called 'demosquitofication' of people's gardens in Réunion.

During the chikungunya epidemic, such 'demosquitofications' were mostly conducted by French metropolitan soldiers working with local DRASS officials. However, the cultural differences between the two often created problems. In addition, the soldiers' ignorance of local practices and concepts of intimate and public spheres within the *jardin creole* led many of the soldiers to disregard the cultural-spatial boundaries of the *kour devant* and the *kour derriere* in their search for mosquito breeding grounds. This in turn could result in friction between the soldiers and local residents. For example, one of my key informants, 72-year-old Gabrielle, who had suffered from chikungunya, risked a substantial fine for denying soldiers access to her *jardin* to conduct a demosquitofication. She also declined their offer of free anti-repellents, as did 41-year old Nathalie: 'I didn't protect myself! There were no products in my house. I walked normally, I dressed normally, and that's it' (Interview, 16 July 2009).

Moreover, Gabrielle and other elderly informants often compared these garden inspections to those that took place during the extensive governmental anti-malaria campaigns on the island during the 1950s (Zettor 2010). With the development of the insecticide DDT in 1939, the fight against malaria accelerated worldwide. In Réunion, anti-malaria work was undertaken by the governmental Prophylactic Services (SDP) and was later taken over by DRASS. Officials from these agencies not only conducted garden inspections similar to those that took place some 50 years later during the chikungunya epidemic, but also sprayed the insides and outsides of peoples' houses with large amounts of pulverised DDT mixed with petrol (ibid.). This method was known locally as 'house painting' and was considered to be highly effective, as it rid people's homes of not only mosquitoes but also of lice, bedbugs, fleas, scabies, and ticks. Despite the island's departmentalisation and full integration with metropolitan France in 1946, French welfare provision and social reforms were only put on the agenda in Réunion during the latter half of the 1960s. Consequently, between 1946 and the late 1960s, the majority of Réunionese suffered from severe poverty and poor sanitary conditions, which contributed to the spread of various vector-borne diseases. As I have discussed in previous papers (Jansen 2013, 2016), even during my fieldwork in 2009 and 2010, Réunionese people commonly associated vector-borne diseases with the work of the SDP. As Raffles (2010) noted, when preventive measures were lacking, living with vectors was a serious and potentially lethal health hazard. For example, in 1948, malaria was responsible for 38 per cent of deaths annually in Réunion (Vaxelaire 2009, 608; Zettor 2010, 18).[6]

Many Réunionese, particularly the elderly, associate these past experiences of unsanitary conditions and malaria with French public health measures. As Gabrielle put it, if one kept one's *jardin* well-kept and clean, consequently there would be no mosquitoes there to transmit disease. Since she fell ill with chikungunya herself, the disease could therefore not be vector-borne since, in her opinion, her *jardin* was immaculate (Jansen 2013, 182). The problem, according to her, was rather the nearby *quartier lavoir* and its Mahoran immigrants. Similarly, Nading (2014) documents the common association of mosquitoes with poor housekeeping in Ciudad Sandino, where residents believe that their neighbours' lack of appropriate domestic cleanliness contributes to the spreading of dengue. Seventy-two-year-old Sylvaine, who lived in a social-housing estate, compared mosquitoes to rats, the classic symbol of the spreading of bubonic plague. To my surprise, several younger informants who had no personal experience with the previous malaria-eradication programmes also claimed that chikungunya was not a vector-borne disease. For example, Nathalie compared chikungunya to 'a passage of polluted air', a bad-air hypothesis of chikungunya infection similar to traditional miasma theories that were also held by many other informants (Jansen 2013).

Miasmatic Disease, Colonial Medicine, and Power Play in Réunion

In miasma theory, which originated in the Middle Ages, diseases were believed to be caused by the presence in the air of 'miasma', a poisonous vapour caused by rotting, foul-smelling organic material. Miasma theory remained the dominant aetiological explanation for disease diffusion well into the nineteenth century when rapid industrialisation and urbanisation in Europe created many poor and unsanitary neighbourhoods that often tended to be focal points of epidemics. Improved housing and sanitation removed bacteria and mosquito breeding grounds.[7]

[6] Malaria was eradicated in Réunion in 1979 (Fontenille et al. 2009, 79).

[7] For example, malaria originates from the Italian words *mala* (bad) and *aria* (air).

Although germ theory largely replaced miasma as an explanation of contagion from around the 1870s, it is still used to make cultural sense of disease by many ordinary people worldwide (Herring and Swedlund 2010; Rosenberg 1992). Certainly, many Réunionese, even today, believe that miasma is the cause of chikungunya. Miasma theory implies exposure to unsanitary environments, in particular noxious mists or vapours from decomposing organic matter near human dwellings, such as the dirty and stagnant water in which mosquitos can breed. Of my 16 informants, 9 claimed that chikungunya was transmitted through foul air. The seven others favoured a mixture of miasma theory and biomedical explanations in which mosquitoes were considered to have first contracted the chikungunya virus from rotting organic material and then spread it to humans (Jansen 2013).

As noted, local people doubted that chikungunya was a vector-borne disease, in large part due to their suspicions about the public health policies of DRASS officials and French soldiers during the 2005–2007 chikungunya epidemic. The discourse surrounding the anti-malaria campaigns of the 1950s and the general political climate on the island at the time also contributed to popular antipathy to the authorities. There was, for example, mounting criticism of the failure of the French government to improve local economic and sanitary conditions. Such sentiments were in part responsible for the 1959 Réunion Communist Party declaration in favour of independence from France. However, such a prospect was anathema to Paris, which considered Réunion to be a key military base in the region (Finch-Boyer 2010). Consequently, France withheld welfare provisions until the island's residents voted against independence in the 1963 election, swayed by local politicians loyal to France, who argued that independence would result in greater poverty, while continued attachment to France would be rewarded by substantial economic, social, and health benefits (Vergès 1999; Jansen 2016).

Following the pro-France vote in the 1963 election, substantial infrastructural changes were finally initiated in housing, schools, roads, and electricity (Vaxelaire 2009), and the same rights to health insurance, social and family allocations, and social housing were provided as to metropolitan French citizens (Finch-Boyer 2013; Vergès 1999). This 'welfare colonialism' continued to ensure that Réunion remained French by making it economically dependent on the metropole (Finch-Boyer 2013). I here contend that 'welfare colonialism' also made Réunion French with regard to disease on the island. Thus, the metropolitan French notion of 'tropicalism', meaning the association between disease and so-called 'exotic' and 'primitive' places and people (Weinstein and Ravi 2008), became current in Réunion where it continued to be reflected in many local illness narratives also during the 2005–2007 chikungunya epidemic. This association was also reinforced by scientific communication of chikungunya. For example, the dust jacket of the popular scientific book '*Le chik, le choc, le cheque*' (The chik[ungunya], the shock, the check) published in 2006 by two locally based medical doctors notes:

> Réunion has painfully reconnected, due to the interference of chikungunya, with its ancestral tropical and African roots. What is this mysterious virus with a cannibal name which bends the spine and eats at the cartilage and pride? (Gaüzère and Aubry 2006)

Moreover, despite the lasting threat of subtropical and vector-borne diseases in all French DOMs in the Indian Ocean and the Caribbean, not until 2006 were chikungunya and dengue included in the French notifiable disease surveillance system (Weinstein and Ravi 2008, 227). These tropical diseases have thus literally not been on Réunion's epidemiological radar. Thus, many Réunionese relate tropical disease, chikungunya included, to the island's colonial past rather than its French present, perceived as the implementation of sanitation and civilisation through welfare colonialism. For example, Gabrielle compared the previous poor health and sanitary conditions of Réunion to what she perceives to be the present state of Mauritius, which she considers to be, in negative terms, highly 'exotic' compared to Réunion.

Conclusion

Réunionese responses to the chikungunya epidemic both challenge past and current public health measures, by employing miasma theory as 'alternative' disease aetiology to vector-borne theories, and simultaneously mirroring colonial racialist health discourses (Winterbottom and Tesfaye 2016). There were, for example, close connections between miasma theory, colonial medicine, and early hygienic approaches to public health in tropical locations (Greene et al. 2013, 50). While European colonists settled in hilltop areas where a vast supply of fresh air kept them above the range of mosquitoes, healthy native-borne populations, who were often considered to be vectors for disease themselves, were relegated to the humid, hot, 'fetid', and densely populated lowlands, often in port cities. In Réunion, not only were local people relegated to the lowlands by colonists, but the mind-set accompanying the spread of vector-borne disease to particular environments also appears to have been adopted by many Réunionese. Moreover, it seems to continue to influence their view on disease. Since vector-borne diseases such as chikungunya have unsanitary connotations in Réunion, mosquitoes were not perceived as responsible for spreading the disease. Instead public health officials and soldiers with so-called accusatory attitudes concerning local domestic cleanliness were considered trespassers in ordinary people's gardens during the epidemic.

References

Archaumbault, J.S. 2016. Taking Love Seriously in Human-plant Relations in Mozambique: Toward an Anthropology of Affective Encounters. *Cultural Anthropology* 31 (2): 244–271.

Arnold, David. 1991. The Indian Ocean as a Disease Zone, 1500–1950. *South Asia* 14 (2): 1–21.

Audifax, M.B. 2006. *Rapport d'Information Assemblée Nationale, no. 3242, Juillet 2006, Déposé par la Commission des Affaires Culturelles, Familiales et Sociales sur l'Épidémie à Virus Chikungunya à la Réunion et à Mayotte.* Unpublished Report.

Bonn, D. 2006. How Did Chikungunya Reach the Indian Ocean? *The Lancet Infectious Diseases* 6 (9): 543.

Castro, A., Y. Khawja, and J. Johnston. 2010. Social Inequalities and Dengue Transmission in Latin America. In *Plagues and Epidemics: Infected Spaces Past and Present*, ed. D.A. Herring and A.C. Swedlund, 231–251. Oxford: Berg.

Conan, M. 1999. From Vernacular Gardens to a Social Anthropology of Gardening. In *Perspectives on Garden Histories*, ed. Michel Conan, 181–205. Washington: Dumbarton Oaks Research Library and Collection Washington, D.C.

Emerson, R.M., R.I. Fretz, and L.L. Shaw. 1995. *Writing Ethnographic Fieldnotes.* Chicago: University of Chicago Press.

Eriksen, T.H. 1998. *Common Denominators: Ethnicity, Nation-building and Compromise in Mauritius.* Oxford: Berg.

Finch-Boyer, H.C. 2010. *Governing Rights in la Réunion: Social Legislation, Landholding, Housing and the Making of France in the Indian Ocean, 1946–2009.* PhD dissertation, University of Michigan.

———. 2013. "The Idea of the Nation was Superior to Race": Transforming Racial Contours and Social Attitudes and Decolonizing the French Empire from la Réunion, 1946–1973. *French Histories Studies* 36 (1): 109–141.

Fontenille, D., C. Lagneau, S. Lecollinet, R. Lefait-Robin, M. Setbon, B. Tirel, and A. Yebakima. 2009. *La Lutte Anti Vectorielle en France/Disease Vector Control in France*. Marseille: IRD (Institut de recherche pour le développement) Éditions.

Gaüzère, B.A., and P. Aubry. 2006. *Le Chik, le Choc, le Chèque: L'Épidémie de Chikungunya à la Réunion 2005–2006 en questions*. Sainte Marie: Azalées Éditions.

Goffman, E. 1992 [1959]. *The Presentation of Self in Everyday Life*. Oslo: Pax Forlag.

Greene, J., M.T. Basilico, H. Kim, and P. Farmer. 2013. Colonial Medicine and Its Legacies. In *Reimagining Global Health: An Introduction*, ed. P. Farmer, J.Y. Kim, A. Kleinman, and M. Basilico, 33–74. Berkeley: University of California Press.

Hasna, M.K. 2003. *Anthropology of the Gardens and Gender in Amazonia and Melanesia*. Unedited paper submitted to the X11 World Forestry Congress, Quebec City.

Herring, D.A., and A.C. Swedlund. 2010. Plagues and Epidemics in Anthropological Perspective. In *Plagues and Epidemics: Infected Spaces Past and Present*, ed. D.A. Herring and A.C. Swedlund, 1–21. Oxford: Berg.

Jansen, K.A. 2012. The Printed Press's Representations of the 2005–2007 Chikungunya Epidemic in Réunion: Political Polemics and (Post)colonial Disease. *Journal of African Media Studies* 4 (2): 227–242.

———. 2013. The 2005–2007 Chikungunya Epidemic in Réunion: Ambiguous Etiologies, Memories and Meaning-making. *Medical Anthropology* 32 (2): 174–189.

———. 2016. Tropical Disease and the Making of France in Réunion. In *Histories of Medicine and Healing in the Indian Ocean World, Volume Two: The Modern Period*, ed. A. Winterbottom and F. Tesfaye, 147–171. New York: Palgrave Macmillan.

Jenkins, V.S. 1994. *The Lawn: A History of an American Obsession*. Washington: Smithsonian Books.

Kirksey, S.E., and S. Helmreich. 2010. The Emergence of Multispecies Ethnography. *Cultural Anthropology* 25 (4): 545–576.

Leyral, P. 2006. Les Autorités Sanitaires de l'Île le Reconnaissent Officiellement: un Enfant est Mort Directement du Chikungunya. *Le Journal de l'île de la Réunion* (February): 14.

Malinowski, B. 1935. *Coral Gardens and Their Magic*. New York: American Books.

Nading, A.M. 2014. *Mosquito Trails: Ecology, Health and the Politics of Entanglement*. Oakland: University of California Press.

Njenga, M.K., L. Nderitu, J.P. Ledermann, A. Ndirangu, C.H.L. Kelly, R. Sang, K. Sergon, R. Breiman, and A.M. Powers. 2008. Tracking Epidemic Chikungunya Virus into the Indian Ocean from East Africa. *Journal of General Virology* 89 (11): 2754–2760.

P. D. 2006. Nouveau Plan de Lutte Contre l'Épidémie de Chikungunya: l'Armée se Mobilise. *Le Quotidien de l'île de la Réunion et de l'Océan Indien* 21 (January): 6.

Payrard, M. 2006. Dispositif de Lutte Anti-vectorielle: les Militaires s'en Vont, les Moustiques Restent. *Le Journal de l'île de la Réunion* 23 May: 14.

———. 2007. Annonce Officielle du Ministre de la Santé: Épidémie de Chikungunya est Terminée. *Le Journal de l'île de la Réunion* 20 April.

Pialoux, G., B.A. Gaüzère, S. Jauréguiberry, and M. Strobel. 2007. Chikungunya, an Epidemic Arbovirus. *The Lancet Infectious Diseases* 7: 319–327.

Raffles, H. 2010. *Insectopedia*. New York: Pantheon Books.

Rallu, J.-L. 2009. Populations et Sociétés: Populations et Développement dans l'Outre-mer de l'Union Européenne. *Bulletin mensuel d'information de l'Institut national d'études démographiques* 456: 1–4.

Raude, J., and M. Setbon. 2009. The Role of Environmental and Individual Factors in the Social Epidemiology of Chikungunya Disease on Mayotte Island. *Health & Place* 15: 689–699.

Renault, P., D. Sissoko, M. Ledrans, V. Pierre, and G. Brücker. 2008. L'Épidémie de Chikungunya à la Réunion et à Mayotte, France, 2005–2006: le Contexte et les Questions de Surveillance et d'Évaluation posées. *Bulletin épidémiologique hebdomaire* 38: 39–40.

Roinsard, N. 2010. Travailler, Chômer, s'Entraider. Discontinuité du Travail et Organisation Sociale à La Réunion. In *Chroniques d'une Autre France: La Réunion. Genres de Vie et Intimités Creoles*, ed. D. Le Gall and N. Roinsard, 21–43. Paris: L'Harmattan.

Rosenberg, C.E. 1992. *Explaining Epidemics and Other Studies in the History of Medicine*. New York: Cambridge University Press.

Taglioni, F., and J.S. Dehecq. 2009. L'Environnement Socio-spatial comme Facteur d'Émergence des Maladies Infectieuses: le Chikungunya dans l'Océan Indien. *EchoGéo* 9: 1–42.

Vaxelaire, D. 2009. *Le Grand Livre de l'Histoire de la Réunion. 1. Des Origines à 1848 et 2. De 1848 à l'an 2000.* Chevagny sur Guye: Collection le grand livre, Orphie.

Vergès, F. 1999. *Monsters and Revolutionaries: Colonial Family Romance and Métissage*. Durham: Duke University Press.

Watin, M. 2008. Polémique, Rumeur et Tension: Aspects de la "Crise" du Chikungunya dans l'Espace Public Médiatique Réunionnais. In *Épidémie et pharmacopée traditionnelle dans l'histoire des îles de l'Océan Indien*, ed. S. Fuma and J. Chan, 241–252. St Denis: Université de la Réunion.

———. 2010. La Médiatisation de l'Épidémie de Chikungunya à Maurice et à La Réunion (2005–2006). In *La santé dans l'Espace Public*, ed. H. Romeyer, 133–149. Rennes: Presses de l'EHESP.

Weinstein, P., and S. Ravi. 2008. The Failure of Colonial "Distancing": Changing Representations of the 2005–06 Chikungunya epidemic in Réunion, France. *Singapore Journal of Tropical Geography* 29: 221–235.

———. 2009. Print Media Representations of an Unusual Health Event: Chikungunya Virus, Risk and Identity on Réunion Island. *Transforming Cultures Ejournal* 4 (2): 144–165.

Winterbottom, A., and F. Tesfaye. 2016. Introduction. In *Histories of Medicine and Healing in the Indian Ocean World, Volume One: The Medieval and Modern Early Period*, ed. Anna Winterbottom and Facile Tesfaye, 1–35. New York: Palgrave Macmillan.

Wolff, É. 1991. *Quartiers de vie: Approche Ethnologique des Populations Défavorisées de l'Île de la Réunion*. Paris: Méridiens Klincksieck.

Zettor, C.M. 2010. Étude Historique de l'Évolution de Pratiques de Lutte Anti-vectorielle à la Réunion. St Pierre, Réunion: Unpublished internal document, *Agence de santé del'Océan Indien* (The Health Agency of the Indian Ocean, ARS).

CHAPTER 11

Inherited Without History? Maldive Fever and Its Aftermath

Eva-Maria Knoll

Although not been recognized by the international media, the population of the Republic of Maldives has had to face a major health dilemma—the world's highest prevalence of beta-thalassaemia, an inherited single-gene disorder that affects the body's ability to create haemoglobin, the red blood cells crucial to providing oxygen to the cell tissue. The most serious form, beta-thalassaemia major, results in severe anaemia in the first months of life and requires lifelong care.

In a Facebook posting, Mariyam, a 25-year-old beta-thalassaemia-major patient, revealed that every 20 days, she has to spend about seven hours in the transfusion clinic receiving donor blood. This procedure started when she was an infant, so she cannot recall her first blood transfusion. While regular blood transfusions have kept Mariyam alive and healthy, there is also the risk that they might poison her because each unit of blood adds critical amounts of iron to her body, which accumulates especially in her organs. To remove this excess iron, she uses a pump to inject iron chelator slowly into her body over a period of about eight hours, five days a week.

E.-M. Knoll (✉)
Institute for Social Anthropology, Austrian Academy of Sciences, Vienna, Austria
e-mail: Eva-Maria.Knoll@oeaw.ac.at

G. Campbell, E.-M. Knoll (eds.), *Disease Dispersion and Impact in the Indian Ocean World*, Palgrave Series in Indian Ocean World Studies, https://doi.org/10.1007/978-3-030-36264-5_11

Mariyam's portrait, which was posted on Facebook by a local self-help group under the rubric "meet our warriors", serves to reveal the everyday struggles and burdens of thalassaemia patients and to demand an "equal chance in life" for them. Unlike healthy Maldivians, *thalassaemics*, as some patients of this blood disorder call themselves, might experience difficulties accessing higher education, engaging in sports, securing a job, being accepted as a marriage partner or founding a family. They are fighting the chronic anaemia that afflicts them, inadequate health care and social stigma. The "warrior portraits" postings serve as role models in this threefold struggle.

Mariyam's portrait was posted on Facebook on 8 May 2016, International Thalassaemia Day. In the Maldives, as in other heavily affected countries, this is a busy day for local NGOs, health professionals, patients, care givers and official bodies such as the Ministry of Health who are involved in organizing public events to raise awareness among Maldivian islanders of the debilitating and life-threatening inherited blood disorder. Symbols of the disorder, such as red haemoglobin discs and red blood drops, are omnipresent (see Pictures 11.1 and 11.2). Catchy pedigrees exemplifying the Mendelian laws of inheritance for single-gene disorders aim to help to translate genetic knowledge into applicable lay knowledge: in every pregnancy, two carriers of a mutated globin gene (heterozygous) have a 25 per

Picture 11.1 International Thalassaemia Day 2015; capital island Male' (Pictures by E. Knoll)

Picture 11.2 International Thalassaemia Day 2015; capital island Male' (Pictures by E. Knoll)

cent chance of conceiving a thalassaemic (homozygous) child, born with two thalassaemia genes, like Mariyam, who are dependent on lifelong biomedical care (cf. Chatjouli 2013). Public outreach events on May 8 seek to encourage people to both undergo genetic carrier testing, with the goal of governing reproductive behaviour in order to avoid further beta-thalassaemia major cases, and donate blood for the recurrent requirements of about 600 thalassaemia patients in the Maldives, a 94,000-square-kilometre Indian Ocean archipelago of 1200 islands.

For the Maldives, as a Small Island Developing State, thalassaemia has far-reaching economic, political and social implications.[1] It is a cost-intensive logistical challenge to provide treatment and prevention services

[1] Thalassaemias are among the most common inherited human diseases. The term thalassaemia refers to a group of blood disorders affecting the production of the two proteins needed to form haemoglobin, Alpha and Beta. The haemoglobin of thalassaemics lacks the ability to bind oxygen. Various haemoglobin gene mutations "run" within Maldivian families: in addition to the majority of beta-thalassaemia carriers, current prevention programmes also screen for alpha-thalassaemia, haemoglobin E and D, and sickle cell carriers. In line with the terminology of the Thalassaemia International Federation (TIF), these haemoglobin gene mutations are covered on the Maldives by the generic term thalassaemia, which I also use in my work.

for its small but widely dispersed population of 340,000, many of whom suffer from inadequate health care and a poor transport infrastructure (Aboobakuru 2014; Firdous 2005). Thalassaemia impacts not only thalassaemic patients but also their parents, kin, spouses and blood donors. Indeed, since every fifth Maldivian is a carrier of the thalassaemia trait, the disorder is perceived not as one affecting an individual but rather as the "joint fate" of a carrier couple potentially "becoming one flesh" (Prainsack and Siegal 2006, 21). Public awareness events help to spread vital information about the disease and to remind all Maldivians that theirs is a genetically high-risk society.

A national thalassaemia prevention programme started in 1992, offering the population screening for carrier identification, the target groups being high-school students and couples who were about to get married. Since this volume is the outcome of a conference organized by the Indian Ocean World Centre at McGill University, it might be of interest that the Maldivian approach to screen students around puberty follows the example of a programme developed at McGill, at Montreal Children's Hospital Research Institute. The Canadian paediatrician and biochemical geneticist, Charles R. Scriver, pioneered adolescent screening in the early 1980s for Quebec's high-risk immigrant communities from the Mediterranean, the Middle East and Southeast Asia, regions where inherited blood disorders are endemic (Scriver et al. 1984; Capua 1998). Scriver reasoned that early knowledge of one's risk status as a carrier would guide reproductive behaviour and help to avoid risky relationships between carriers which, in turn, would reduce the number of affected newborns. The Montreal high-risk minorities and the entire Maldivian at-risk population are approximately comparable in size. However, while the Quebec screening programme was aimed at just 2.78 per cent of potential thalassaemia carriers among Montreal's population residing in consolidated ethnic neighbourhoods (Mitchell et al. 1996), the Maldivian carrier screening programme caters for a small and dispersed island population of which some 20 per cent are carriers. In consequence, improving thalassaemia treatment and reducing the number of new thalassaemia major patients (at the time of writing over 20 each year) are top health priorities in the Maldives (MoH 2015, 20).

Biomedicine and genetics, however, reveal only how individuals became sick, not *why* Maldivians are much more susceptible to thalassaemia than other people. During the five years that I studied thalassaemia in the Maldives, I was time and again addressed with the nagging "why us" question. As a male thalassaemic from a southern atoll aptly put it: "These islands are called paradise. How come that we [paradise dwellers] have to check before we fall in love?"

There exists no historical study of the socio-economic impact of inherited chronic anaemia in the Maldives. By contrast, thalassaemia in South India and amongst Pakistani Muslims has been traced historically to consanguineous marriages (Bashyam et al. 2004, 412; Ahmad et al. 2000, 24–25), and the sickle cell trait in the Caribbean to the slave trade (Dyson et al. 2016, 623–627). In the Maldives, however, no comparable narrative connects the abstract genetic pedigrees shown at awareness events and genetic counselling sessions with any historical legacy.[2] As an inherited recessive single-gene disorder, thalassaemia is passed down through the generations. Of course, the ancestors of today's carriers also mated and conceived thalassaemic offspring. At probably between 6 and 24 months of age, a homozygous thalassaemic infant would become severely ill and die, largely unnoticed because of a traditionally high infant and child mortality rate (cf. Hedrick 2011, 291). This was particularly the case as the Maldives suffered from endemic malaria which causes symptoms similar to thalassaemia in infants, notably severe anaemia and enlarged spleens (Weatherall 2010, 18). It is probably due to this that thalassaemia and patients suffering from it are not detectable in written sources before the field of genetics reached this remote island corner of the world in the 1990s.

How is it that one of the smallest and least populous countries on the planet has the world's highest rate of beta-thalassaemia carriers and sufferers? And why is this genetic shadow in the island paradise largely unknown? This chapter seeks to provide some historic context to the aetiology and epidemiology of the exceptional prevalence of thalassaemia in the Maldives through "re-fusing" (Dyson et al. 2016) abstract genetic information and disease impact with the history of the islands. It first establishes a link between thalassaemia and malaria; second, it reviews historical reports on the legendary Maldive fever; and finally it analyses those reports in terms of the environmental and socio-cultural context.

[2] The majority of my research interlocutors had no knowledge of the aetiology of thalassaemia. Only a few among them were afraid the blood disorder might have resulted from the multilayered degrees of kin relations within the small island communities. However, in contrast to other Muslim societies, parallel cousin marriage is not practised in the Maldives (Firdous 2005, 132). A few of my interlocutors speculatively assumed that the Portuguese, during a short-lived colonial domination of the Maldives from 1558 to 1573, had brought "bad genes" to the islands—linking the contemporary thalassaemia burden to a painfully remembered historic period and to the national pride of eventually defeating the odious Catholic colonizer (Knoll 2018a).

Vector-Borne Disease and Selective Advantage: The Malaria Hypothesis

Detroit paediatricians Thomas B. Cooley and Pearl Lee (1927) were the first to describe the clinical manifestation of the thalassaemia syndrome, also known as Cooley's anaemia. Recognizing clusters of the syndrome in certain families or populations, they assumed it to be an inherited disease. In 1945 this assumption was confirmed, and thalassaemia was recognized as a Mendelian autosomal recessive blood disorder. Given high rates of carriers of thalassaemia and other lethal haemoglobin disorders in certain populations around the globe, it was assumed that since these mutated genes are plentiful and of high frequency, but had not wiped out the affected populations, they must represent an added value in terms of selective adaptations. British scientist John Haldane (1892–1964), one of the founders of population genetics, noted that the worldwide distribution of haemoglobin mutations generally coincides with the historic distribution of the vector-borne disease malaria. Due to its high mortality rate and widespread impact, malaria is recognized as the "strongest evolutionary selective force in recent human history" (Hedrick 2011, cf. Green and Jones, this volume). In 1949, Haldane suggested "the possibility that the heterozygote [carrier] is fitter than the normal" and "a possible mechanism" for this increased evolutionary fitness:

> The corpuscles of the anaemic heterozygotes are smaller than normal, and more resistant to hypotonic solutions. It is at least conceivable that they are also more resistant to attacks by the sporozoa which cause malaria. (Haldane 1949, 270)

The "idea that genetic variants in humans may confer resistance to malaria" might explain the high level of carriers in particular populations (Hedrick 2011, 284). This "balanced polymorphism", commonly referred to as the "malaria hypothesis", postulates evolutionary

> trade-offs whereby one benefit outweighs another cost. … Having one sickle [or thalassaemia] gene confers substantial protection [against malaria mortality and severe malaria symptoms], and this advantage ensures a high frequency of these genes, despite the fact that those with two copies of the gene develop a disease. (Chiou 2016, 43)

As non-carriers of sickle cell or thalassaemia genes more often fell victim to the malaria parasite, carriers of either of those genes have, over the centuries, come to form significant portions of the population of malarial areas. Finally, improvements in molecular research techniques indicated strong geographical correlations between haemoglobin variant frequency and malaria endemicity, confirming Haldane's hypothesis of gene-based responses to malaria (Piel et al. 2010; Weatherall 1997).

Consequently, this investigation will focus on fever as the prominent symptom of malaria in historical sources. In this regard, the Maldive Islands had such a reputation that a fever is named after the archipelago.

Maldive Fever

There exist few histories of the Maldives, and sources on its past are scant (e.g. Picture 11.3), probably due to the archipelago's remote location and its comparatively small population and insignificant natural

Picture 11.3 Maldive Island 1957–1958. (Picture by Irenäus Eibl-Eibesfeldt)

resources (Knoll 2018a).[3] Given this, it is challenging to trace the history of health and disease in the islands. The historical (written) sources for this article are predominately based on eye-witness accounts of the health hazards there. Reports on the "severe and very troublesome" sickness "known through all the Indies under the name of Maldive fever" (Pyrard 1887, 82–83) span about six centuries, from early Arab travel to the heyday of European expansion. Building upon an earlier brief discussion on the subject (Knoll 2018b, 319–321), I will try to dig deeper by consulting additional sources and, subsequently, by assessing these historical voices in dialogue with their environmental and socio-cultural contexts.

Carl Wilhelm Rosset (1851–1923), a German adventurer and collector, visited the British island colonies of the Maldives, Laccadives and Ceylon in the course of an expedition between 1884 and 1886. Approaching the Maldive archipelago from the sea, Rosset reports this first encounter:

> The panorama, which was now spread out before us, was beautiful in the extreme. The low shore, marked by the thin white line of the beach, was covered to the height of about seven feet with a thick growth of jungle, above which waved the graceful heads of thousands of coconut trees, to which the slight breeze then blowing imparted a scarcely perceptible motion.
>
> As I leant over the bulwarks, admiring the scene, I suddenly became aware of a painfully pestilential odour, which at once dissipated the romantic thoughts, which the beauty of the scene had conjured up. This was the much dreaded fever-laden breath of the lagoons, the cause of the deadly Maldive fever.
>
> This stench is due to a peculiarity in the atolls, or clusters of islands and reefs which constitute the Maldive group. Most atolls are formed of a circle of islands, connected by reefs, which enclose a large tract of water, or lagoon…
>
> Almost all Europeans who, by different shipwrecks, have landed on these islands, have died there from fever. During my stay in the islands, up to 60% of the natives were sick with fever, of course due to the unhealthy season. (Rosset 1886/1887)

As a "gruff Victorian explorer-adventurer who exposes himself to all sorts of dangers and discomforts in the name of a higher (national [and scientific]) mission" (Pratt 1986, 39), Rosset wove together a personal, particularized narrative with factual and generalized descriptions. The "narration-description duality" (ibid., 35) of arrival stories that inserts the

[3] Remarkable work in locating texts and sources on Maldivian history in libraries and archives around the world was carried out by the Swedish history researcher Lars Vilgon, compiled in a bibliography and nine "Maldive Odd History" volumes.

authority of personal experience in the text was the conventional style for Victorian travel accounts. Furthermore, in the above passage, Rosset provides a reversal of the discursive legacy of utopian accounts of arrival on idealized tropical islands. With the "anti-utopian instance" (ibid., 41) of pestilential odours and malignant fever, Rosset's Maldives becomes a fallen paradise, its environment an obstacle to European penetration.[4]

Scouring Rosset's and other historic reports for epidemiological indications makes it possible to identify some general characteristics of Maldive fever. First, there are seasonal fluctuations in its occurrence that have contemporary echoes. Rosset arrived in the archipelago during the "unhealthy season" which he held responsible for 60 per cent of the locals being sick with fever. During the rainy and stormy, low-tourism season of the southwest monsoon (*hulhangu moosun*), a considerable percentage of *Dhivehin* (Maldive islanders) still catch colds, or come down with a severe viral infection such as mosquito-borne dengue fever.[5]

Second, Maldive fever was considered inescapable due to a climate notoriously unhealthy for Europeans and hardly salubrious even for the indigenous population (Bell 1882, 6–8). The fourteenth-century Muslim traveller Ibn Battuta (1304–1368/1369) (cited in Vilgon 2001a, 41) and François Pyrard de Laval (1578–1621), a chronicler in a French merchant expedition to the East Indies who got shipwrecked on a Maldivian reef in 1602, were convinced that every foreigner visiting the archipelago "must inevitably catch the fever" (Pyrard 1887, 85).[6]

In 1723, the Dutch dispatched two ships to the islands to acquire cowry shells. An "extremely uncooperative court", however, "kept (them) dangling" for several weeks in the insalubrious climate before they were allowed to trade products and buy cowries. "One after the other felled by

[4] I am grateful to Burkhard Schnepel for drawing my attention to the writing conventions of Rosset's report.

[5] To name a more recent example: in March 2017, in the transition period from dry and rainy season, an H1N1 (commonly called swine flu) epidemic added to the seasonal fever spike caused by respiratory infections, influenza and dengue. When more than 5000 cases were registered in a single week, the number of confirmed cases reached 51 and the epidemic claimed a first victim, public schools were closed in order to prevent further spreading.

[6] In his study of the malaria epidemic that hit Mauritius in 1867, Nobel Laureate Sir Ronald Ross revealed what a newly introduced virus can do to an immunologically naïve population. The IOW chikungunya epidemic 2005–2007 with some 1.9 million cases illustrates this as well (see Alpers, this volume). While in these two disease distribution instances, new viruses hit naïve populations, in the Maldivian case it was the naïve immune system of the visitors that got hit by the endemic Maldive fever.

fever, the Dutch crew went down like ninepins and could only wait and see. Before the ships could return, both commanders had died and been buried" (Raben 1996, 51). The Archaeological Commissioner of Ceylon, H.C.P. Bell (1851–1937), visited the Maldives three times between 1879 and 1922. In October and November 1922, "no less than three hundred (300) victims, Noble and plebian [sic!] alike, perished from this scourge" (Bell 2002, 6), that is, 5 per cent of the capital's population of 6127 (Census from 1921; ibid., 14). Bell described the "dread 'Máldive gift-fever' a type of deadly 'Influenza' notorious for centuries past" which even strikes people long after they have left the islands. Since all members of his mission "have more or less suffered from half-yearly, and even more frequent attacks of this dangerous form of ailment, since their return to Ceylon", Bell suspected that the "seeds" of the febrile disease "must have been sown" (ibid., 6) in the Maldives, although its malign effects only appeared after the mission had left the islands.

Third, Maldive fever was linked to "an obstruction and inflammation of the spleen". Pyrard (1887, 84) reports further, "this spleen disease is very common among them [the Maldivians] and they all have it rather large". Fourth, historic epidemiological considerations centred on two parameters: bad air and bad water, which combined notably in the form of the foul-smelling stagnant waters of lagoons, swamps and marshes. Dr David Campbell, surgeon on the East India Company ship *Benares* during her survey of the islands in 1835, assumed, for example, that the stagnant waters, interspersed with decaying jungle vegetation, were "generating a poison, or 'Malaria' [*Mala aria*—mediaeval Italian for 'bad air'] productive of those fevers, when coming in contact with the human body" (cited in Vilgon 2001b/ IV, 77, 81).

In equatorial latitudes, increased body temperature or "fever" can have many causes, including insect-transmitted parasites. Mosquitoes, in particular, can transmit a multiplicity of diseases. Filariasis, for example, commonly known as elephantiasis, was endemic and particularly pronounced in the Maldives (Iyengar 1952). Scrub typhus, also endemic, is transmitted through larval and adult mites. In 1941, a battalion of 1059 Royal Marines on the island of Gan in Addu Atoll were badly affected by scrub typhus fever while clearing undergrowth in order to construct an airfield (Royal Naval Medical Service 1954, 218–220). Most scholars, however, assume the malady that was referred to over six centuries as "Maldive fever" was malaria (e.g. Maloney 1980, 133, 398–399; WHO 2016, 1) or a combination of particularly virulent malaria and common gastrointestinal infections such as Shigella, cholera or dysentery (Kläy 1986, 89).

The Socio-cultural and Economic Environment of Mosquito Breeding Grounds

Given eye-witness reports of the dense distribution, inescapability and severity of the fevers in the Maldives, we can assume that there must have been vast numbers of mosquitoes and, despite a general scarcity of surface-water resources, multiple mosquito breeding grounds. An early entomological research project based on collections from five atolls alone recorded 15 different species of mosquito, 2 of them of the genus *Anopheles* (Iyengar and Menon 1955). Mosquitoes breed in the few brackish and freshwater lakes of the Maldives, in the brackish waters of its mangroves, in swampy taro fields (see Picture 11.4), salt-marsh grass meadows, rainwater tanks, wells and bathing tanks. There are no springs, rivers or streams in the Maldives. The country's natural freshwater resources comprise a few freshwater lakes and, in the larger islands, groundwater in basal aquifers extend-

Picture 11.4 Taro field on Fuamulak, Southern Maldives. (Picture by E. Knoll)

ing below sea level in the form of a thin freshwater lens. The islands are dependent on rainfall and vulnerable to pollution from human activities on the island surface and to saline intrusion by the surrounding sea. These limited and fragile natural freshwater reserves, and collected rainwater, provided the only drinking water resource (Ibrahim et al. n.d.) and, unintentionally, mosquito breeding grounds.

A *valhu* (well) or *veyo* (bathing tank) existed in every fenced compound that housed one extended family. In the 1970s, Velimirovic and Clarke (1975, 505) estimated that there were some 20,000 wells in the archipelago. There exist three kinds of *veyo* characterized by the traditional water management system. The first is the family *veyo* (sometimes called a step well). In the diary of his 1920 visit to Male', Bell described a *veyo* in the compound of an affluent and prominent family:

> In the back garden is built a spacious bath-room (23 ft. by 18 ft. [7 x 5.5 m]), in the centre of which has been sunk a large circular swimming-bath of cement (13 ft. [4 m] in diameter), with a broad flight of steps, and a wide gangway round—a most luxurious annexe, always kept filled with clear water. (Bell 1921, 51)

The second is the community *veyo*, which during his visits in 1879, 1920 and 1922, Bell described as slab-built bathing tanks attached to mosques. He considered those in the densely populated capital island of Male' to be health hazards that "are emptied but once a year, and naturally stagnate to a state which beggars description" (Bell 2002, 62):

> Some of these bathing-places—*infandum renovare dolorem*—are allowed to lapse into veritable cess-pools, absolutely green to the eye and most aggressive olfactorily, from the human slime and filth added month by month with but a single 'cleansing' during one solitary day a year. (diary 26 April 1922 as cited in ibid.; cf. Ellen Kattner's description of the annual cooperative cleaning on Maliku (Minicoy) 2007, 165)

During Bell's 1922 visit, two large rainwater storage tanks were built and another two were planned to provide the city with drinking water (Bell 2002, 62). Today no community *veyos* or traditional rainwater storage tanks remain in the capital, which now depends on desalinated seawater and bottled water. Additionally, almost all family *veyos* have been filled. The third kind of bathing tank comprises largely unattended community *veyos* that exist on islands outside Male' and are generally believed to date

Picture 11.5 Ancient *veyo* on Fuamulak, Southern Maldives. (Picture by E. Knoll)

back to pre-Islamic, Buddhist days (see Picture 11.5). Scholars link the impressive traditional water management system of the Maldives to the water architecture that has been acknowledged as a remarkable cultural feature of the wider south and southeast Indian Ocean region. Bell linked the *veyo* in Male' as "strongly reminiscent of the Buddhist *pokuṇu* of Anuradhapura" (Forbes 1987, 282) to the ancient hydraulic societies of Ceylon. Kattner (2007, 151–154) describes similar rectangular community tanks, called *bodu valu*—big ponds—on Maliku (Minicoy) with a surface of 10 to 100 m^2 as "the enigmatic aspect of the traditional water management system" and links them to the elaborated water infrastructure of the Harappa culture. Maliku is the southernmost island of the Indian Union Territory of Lakshadweep but belongs culturally, linguistically and historically to the Maldives.

In addition to these natural and man-made breeding grounds, the regular heavy tropical rainfall throughout the year, which intensifies during the south-east monsoon, results in huge and persistent freshwater puddles—despite the generally high permeability of the sandy soil. The "meteorological conditions are therefore optimum for the activities of the anopheles vectors and consequent transmission of malaria throughout the year", stated WHO Officer, J.R.M. Schepens (1981, 2), in a Malaria Control Report. Today, abandoned cisterns, roof gutters, construction sites and trash—especially plastic waste and tyres—provide ample new mosquito breeding grounds.

Mosquito Bionomics and Maldivian Culture

Interactions between breeding-ground environments, human and animal hosts, mosquito vectors and malarial parasites are complex. There were, for example, close connections between miasma theory, colonial medicine and early hygienic approaches to public health in tropical locations (Greene et al. 2013, 45). Diverging appraisal of wetlands thus can give us valuable epidemiological insights into the history of thalassaemia in the Maldives. Before the advent of the germ theory by the end of the nineteenth century, Europeans such as Rosset viewed stagnant waters and swamps not only as physical barriers to movement and settlement but also as sources of dangerous miasmas, that is noxious mists responsible for the spread of disease. Draining swamps and transforming them into cultivated land was the acknowledged European way to eradicate miasmic gases (Packard 2007, 13). This Eurocentric perspective, however, misses the ecological and economical value of the Maldive wetlands, their intimate interface with Maldivian culture and heritage, and their entanglement in Indian Ocean trade networks (Webb 1988).

The often flooded cultivated meadows of salt-marsh grass provided Maldivians with *hau*, the raw material for *kunaa*, woven grass mats. In many a European museum, such *Cyperaceae* mats (see Picture 11.6) are portrayed as a highlight of Maldivian material culture (Forbes and Ali 1980a; Kläy 1986, 122–135; Ottovar and Munch-Petersen 1980, 56–75). Mangroves also serve as habitats for numerous varieties of bird and fish, and help control flooding and prevent coastal erosion. Further, they have provided the islanders with firewood, timber for boat building, and food. During the annual Eid celebrations, Dhivehin from northern parts of the archipelago apply *mashi*, a type of clay found in the mangroves, to their bodies and faces and parade as *mashi maali* (clay monsters) on the streets.

Picture 11.6 Woven grass mat, collection Carl Wilhelm Rosset 1886, Copyright: KHM-Museumsverband, Weltmuseum Vienna

Mangroves and stagnant lagoons were critically important to coir-rope and cowry production. The coconut husks used in *roanu* (coir-rope) production were softened by being soaked for months in husk-pits in the mangroves or in the waters of the intertidal zone (cf. Adam 2016). Maldivian *roanu* was valued in Indian Ocean sailing "for its light colour, fineness and strength" (Hockly 1935, 102). It was used in the local and regional production of sewn boats and was in high demand from Arab and Gujarati traders and the Portuguese who valued it as durable rigging or cordage (Forbes 1981, 81; Villiers 1992). In addition, cowry shells were collected from coconut palm leaves placed in a lagoon, or picked out from lagoon debris (Hogendorn and Johnson 1986, 80–82). Indeed, the Maldives constituted the major source of cowry shells in the Indian Ocean World (IOW) (ibid., 34), and their wetlands a kind of central IOW cowry mint.

Genetic, Environmental and Biosocial Encounters

Since malaria parasites thrive throughout the year in tropical and subtropical regions, the peoples of the IOW are particularly burdened with evolutionary genetic responses in form of various inherited blood disorders—among them some 200 variants of β-thalassaemia with carrier frequencies of between 5 and 20 per cent (Hedrick 2011, 291). Of the WHO demarked areas of the world, 23 per cent of the annual global pregnancies that require risk assessment due to prevailing inherited blood disorders occur in the South East Asian Region (SEAR), second only to Africa with 58 per cent.[7] The high 19 per cent thalassaemia carrier rate in the Maldivian population (Firdous et al. 2011) illustrates the lasting impact of malaria and malaria adaptation. Consequently, I have argued (Knoll 2018b, 336–337) that the severe and durable impact of malaria and malaria adaptation on people and societies should be added to the constitutive "deep structure" of the IOW that historian Michael Pearson (2003, 13–25) has identified in climate, monsoon winds, tides and currents.

Beyond the local specificity of the Maldivian case, the study of inherited blood disorders opens up a methodological pathway for the investigation of disease dispersion. The diversity, frequency and distribution of contemporary and ancient molecular spectra of haemoglobinopathies allow us to draw conclusions on disease origin and dispersion as much as it illuminates the migratory past of human beings. Yet population genetics of haemoglobinopathies is still a young field with a strong focus on the Mediterranean basin. Sardinia, an island that alongside Cyprus has one of Europe's highest beta-thalassaemia rates, provides evidence of an early historical thalassaemia case. The cod39 single-point mutation, responsible for more than 95 per cent of current beta-thalassaemia cases in Sardinia, was, through analysis of ancient DNA (aDNA), detected in skeletal remains dating back 2000 years (Viganó et al. 2017). The skeleton was that of an asymptomatic adult male carrier of beta-thalassaemia, presumably of mixed Sardinian and Punic descent, and confirmed the continuing presence of malaria as a selective pressure—perpetuating the cod39 mutation in the island population.

[7] See the global epidemiological estimates for haemoglobin disorders in Modell's *Haemoglobinopathologist's Almanac* 2008 (http://www.modell-almanac.net, consulted Sept. 17, 2016).

Less is known about the distribution history of genetic diseases in the IOW's low-income countries which are highly burdened by these inherited blood disorders (e.g. Hewitt et al. 1996). In the first study of its kind in the archipelago, Furuumi et al. (1998) traced the genetic ancestry of the most prevalent thalassaemia mutations in 41 Maldivian thalassaemics. The study detected four different mutations, three of which account for 99 per cent of the beta-thalassaemia chromosomes. Moreover, this Maldivian spectrum of beta-thalassemia gene mutations differs from those found in neighbouring countries. Reflecting the IOW maritime crossroads position of the archipelago, it points to multiethnic linkages with various regions of the IOW. The globin gene causative mutation IVS-I-5 (G>C) is found in Asian Indians, Southeast Asians, Melanesians and Pakistanis, while the codon 30 (AGG>ACG) mutation probably originated in the Middle East, and IVS-I-1 (G>A) probably derived from Portugal and Algeria.

Thus the current Maldivian thalassaemia carrier rate results from a distinct malarial environment and human activities that link the islands to various other IOW regions. The strategic geographic location of the island chain at the Indian Ocean crossroads (Knoll 2018a), their close proximity to the Malabar coast and the seafaring history of the Dhivehin have impacted the gene pool of the population of the Maldives (Mohamed 2008; Pijpe et al. 2013). The Dhivehin sailed as far west as the East African Coast (Forbes and Ali 1980b) and as far east as China (Ptak 1987, 680–681). Settlement patterns and migration flows link the Dhivehin to the Indian subcontinent and Sri Lanka. Their mid-oceanic position involved them in transregional religious and trade networks that in pre-Islamic times included Hindu and Buddhist networks and later, Arabian and Islamic networks. This involvement intensified when the Maldives converted to the Muslim faith in 1153 (Forbes 1981; Maloney 1980, 98–134). Colonial encounters with the Portuguese (Fitzler 1936), Dutch and British probably also left genetic traces.

Further biosocial factors, such as genetic founder effects and marriage patterns, may help to explain the uneven distribution of thalassaemia over the 20 administrative atolls in the archipelago.[8] Local marriage patterns, such as island endogamy, the polygamy practised by the wealthier strata of society and notably high divorce and remarriage rates (Maloney 1980, 309, 336–337, 343–348; Rosset 1887, 168–169), consolidated carrier

[8] Internal Statistic of Maldivian Blood Services (MBS), May 2015.

intensity in certain atolls and islands. Ibn Battuta even reported a form of "temporary marriage" that reflected the long stays enforced on sailing ships that had to await a change in the monsoons in order to return to their home ports (Dunn 2005, 237).

People living at lower altitudes experience generally higher frequencies of beta-thalassaemia variation than those living at higher altitudes, and mosquito vectorss are zoophilic rather than anthropophilic (Hedrick 2011, 285). The geographical environments of the small, low-lying Maldive coral islands, however, were of certain limitations and thus also affected over the *longue durée* the contemporary health situation. The highest point of the Maldives is a mere two metres above sea level, while the only endothermic animals whose blood mosquitoes preferred to that of humans were the endemic giant fruit bat *Pteropus giganteus ariel* (O'Brien 2011), and rats and mice that arrived later in the archipelago. Only 30 of the 187 inhabited Maldive islands are more than 3 km^2 in size, leaving little room for human settlements in locations that were a safe distance from breeding grounds. In short, the island environment did not allow for obvious human responses to mosquito attacks such as settling in higher altitudes or far from wetlands. Thus, in contrast to other parts of the IOW where abundant wild and domesticated animal species formed alternative mosquito prey, the Dhivehin were almost helplessly exposed to these tiny blood-sucking insects. Correspondingly, within the Indian Ocean "disease zone" (Arnold 1991), small tropical island environments become recognized as dispersion hubs for emerging arboviruses such as dengue, chikungunya and Zika (Cao-Lormeau 2016).

Cultural Dimensions: Putting "Maldive Fever" into Perspective

As mentioned before, there were close connections between miasma theory, colonial medicine and early hygienic approaches to public health in tropical locations (Greene et al. 2013, 50). Accordingly, miasma theories dominated the fever aetiology of European travellers approaching the Maldives. The idea that fatal fever was caused by *mal'aria*—by breathing noxious marsh air, or miasma—was rooted in the humoral theory of Greek physicians Hippocrates (460–370 BCE) and Galen (130–210 CE).[9]

[9] The observation of the coincidence of wetlands or rainy periods, mosquitoes and malarial fever is, of course, not a privilege of the Western world. In the first millennium BCE, Indian

However, James Webb (2009, 62) argues that malaria was largely a lowland phenomenon in the Mediterranean basin where it constituted the only mosquito-borne disease. The association of fevers with low-lying swamps thus is "deeply anchored" in the common European experience (ibid.). In the 1880s, when humoral theory was replaced by germ theory, the French military physician Alphonse Laveran (1845–1922) investigated the blood of Algerian "marsh fever" patients under the microscope and identified protozoal parasites as the fever's cause. The subsequent discoveries, notably of the life cycle of the *Plasmodium* parasite made by Ronald Ross (1857–1932) in 1897, and of acquired immunity to *Plasmodium* parasites and the effectiveness of quinine made by Robert Koch (1843–1910) in 1897, as well as Giovanni Battista Grassi's (1854–1925) demonstration of the complete route of human malaria transmission through the identification of the female *Anopheles* mosquito as the malaria vector in 1898, put an end to miasma theories and cleared the way for new, microparasite- and vector-focused approaches to treating and controlling malaria. In 1899, the first attempt to control malaria by eradicating transmitting *Anopheles* mosquitos was carried out in Freetown, Sierra Leone (ibid., 127–130; cf. Packard 2007).

The question remains, how historically Dhivehin made sense of the intense attacks of fever in their islands. While Westerners tended to categorize the disease in terms of fever rhythms and wetland aetiologies, in the Maldives fever was mentally conceived of as a comprehensive term linked to an emic aetiology. In addition to traditional Maldivian folk medicine (*Dhivehi beys*) in which water mixed with ground pepper was prescribed to treat fever (Maloney 1980, 398), disease and healing were strongly associated with the supernatural world. As mentioned above, the small, discrete and low-lying island habitat provided no ecological restriction to malaria and did not allow the social adaptations such as those in Sardinia where people moved to higher-altitude settlements to avoid fever (Brown 2016). By contrast, Maldivian social responses to fever involved recourse to the spheres of superstition and folk belief.

Maldivians believed that disease, including high fever, and other afflictions were caused by *dhevi*—mysterious beings such as spirits, demons, ghosts and jinnis. *Dhevi* were invisible or visible and interfered with humans in harmful or helpful ways (Maniku 1988). *Baburu-kissadhevi*, for

and Chinese authors reported their observations and complex views on fever causation (see the overview in Webb 2009, 58–63).

example, molested women and produced weakness, vomiting and high fever. *Gaskolhu-handi*, also known as *Gaskele Jinni* in the southern atolls, caused in both sexes a kind of fever associated with typhoid. *Kaduranin*, the queen *dhevi* of the sea, brought high fever and shivering, causing people to hallucinate. *Mulhadhevi* inhabited graveyards and caused swellings, stomach pain, yellow eyes and the urge to urinate, as well as fever in the midmorning when the rising sun heated up the graves. *Odivaru Ressi* possessed seafarers provoking amongst them continued fever, cough, itching and excessive thirst. *Ummusubyaanu*, also known as *Kuddhingebiru*, a blue-eyed blue-clothed female *dhevi*, caused a fever in children characterized by abrupt fluctuations of body temperature. The magical healing ritual for the *Ummusubyaanu* fever included the healer jumping with the afflicted child into a bathing tank (ibid., 15–37). *Dhevi*-induced fever could only be cured by *fanditha*—magical practices that were part of syncretistic activities "incorporating ancient local charms along with Koranic verses" (Victor 2014, 382; cf. Maloney 1980, 242–265).

The remarkably rich and comprehensive supernatural world of the *Dhivehin* also included blood-sucking mystical beings (Victor 2014, 382; Romero-Frias 2003, 113–129, 2012). The majority of *dhevi* were active at twilight or during the night. Traditional houses had neither windows nor indoor fires lest they attract malevolent forces, while small oil lamps were kept burning inside mosques and houses all night in order to ward off such forces. If a lamp went out, leaving the place in darkness, it was considered a very inauspicious omen (Maloney 1980, 244–245; Victor 2014, 381; Romero-Frias 2003, 53, 61, 221; Maniku 1988, 37).

The links between fever and other diseases and the supernatural world of the *Dhivehin* deserve more in-depth research. However, it may be safe to hypothesize that if we are ready to attribute a certain "common sense" (Geertz 1983) to European miasma theories, and acknowledge reasoning in analogical dimensions (Lévi-Strauss 1966) between bad smelling swamps and disease causation, then the Maldivian social response to overwhelming fevers deserves similarly respectful consideration. If the Western mind could "smell" the danger in the stagnant waters of the swamps, the Maldivian mind was sufficiently alert to fear something unknown out there in the dark. Villingili, for example, which in 1922 Bell (2002, 7–8) described as an island that "lives well up to its name *Madiri Viligili*" (mosquito island), some 50 years after Bell's comment continued to be "avoided because it was believed to have a thousand *jinnis*" (Maloney 1980, 245; emphasis in original).

The Eradication of Maldive Fever

Given that for some 600 years reports had circulated of the Maldive Fever and its deadly nature, the first attempts at malaria eradication in the islands came comparatively late. The British Crown colony of Ceylon, to the east of the Maldives, was the first Asian country to start a nationwide malaria control programme in 1946, one that involved spraying all homes with DDT (dichlorodiphenyltrichloroethane) (Johnson 1966, 476). Although the Maldives was then a British protectorate that was politically and economically closely tied to Ceylon, and was well known to be plagued with fever, it was excluded from the DDT programme. The World Health Organization (WHO), founded in 1948, put malaria high on its international agenda, and in 1953, with US support, sponsored a country-wide control programme in India which subsequently became a world leader in anti-malaria activities. The same year (1953), the result of a 1951 survey of filariasis in four southern atolls of the Maldives was published. It revealed high spleen rates and endemic malaria emanating from three species of malaria parasites: *falciparum*, *vivax* and *malariae* (Iyengar et al. 1953). This report was the first scientific identification of a malarial problem that had plagued the islands for centuries. However, only in 1965 was a specialized and focused investigation of malaria in the Maldives launched—the same year that the WHO announced that 58 per cent of the 879 million Asians living in malarial areas were "almost free of the threat" (Johnson 1966, 477). The 1965 WHO survey of the Maldives was initially limited to the capital, Male', and other islands of Male' Atoll. It found that an "alarming" 15 to 60 per cent of the population on various islands in this atoll had palpably enlarged spleens due to malarial infection (WHO 2016, 9). An official malaria control programme started on 23 May 1966 (ibid.; Schepens 1981), the year after the Maldives obtained independence, 11 years after the WHO launched the Global Malaria Eradication Programme and just 3 years before the WHO abandoned the goal of global malaria eradication.

The Maldivian programme started with one cycle of spraying indoor residences with 2 g/m^2 of DDT (Velimirovic and Clarke 1975, 504). It continued with larval surveys and the limitation of potential breeding sites, blood tests and mass drug administration of chloroquine and primaquine to every *Dhivehin* for five days. The hospital vessel *Golden Ray* carried the malaria team through the scattered islands of the archipelago. Due to the lack of health facilities at the time, heads of island communities and family health workers were trained in the detection, treatment and follow-up care of malaria patients. Also, representatives from local communities were recruited to assist in the control of potential breeding sites. Larvivorous fish, or kerosene and diesel, and later the larvicide Abate (Temephos) and *Bacillus thuringiensis Israelensis* (BTI) were introduced into freshwater tanks and cisterns to counter mosquito larvae. Trained men sprayed the roughly 200 inhabited and neighbouring uninhabited islands of the archipelago (WHO 2016).

After the last campaign in 1974, malarial morbidity dropped sharply, the *Anopheles* mosquito was considered to have "virtually disappeared" (Velimirovic and Clarke 1975, 504) and "the distended bellies and anaemic eyes of the children became a thing of the past" (Yoosuf 2014, 10). In 1984, anti-malaria and lymphatic filariasis teams also reached the remoter corners of the archipelago and in 2000 merged to form the Vector-Borne Disease Control Unit. The last case of *P. falciparum* was reported in Haa-Alif Atoll in 1975, the last *P. vivax* case in Baa Atoll in 1984 (WHO 2016, 13).

Malaria eradication played a key role in the transformation of the Maldives from one of the poorest to one of the most economically well-off countries in the WHO South East Asia Region (SEAR). It is no coincidence that the advent of the Maldivian tourism industry emerged within the same time frame. Malaria re-importation, transmission and re-establishment, however, remain a looming threat, as does the spread of emerging vector-borne diseases such as dengue and chikungunya fever.

Conclusions

The *Dhivehin* were vulnerable to the three malarial parasites *Plasmodium falciparum*, *P. vivax* and *P. malariae*. About 50 per cent of all WHO documented infections in the Maldives were caused by *P. vivax*. *Anopheles tessellatus* and *A. subpictus* were identified as the primary malaria-carrying mosquitoes (WHO 2016, 9; Schepens 1981). Initially, the archipelago's small and scattered island communities were difficult to logistically integrate in an eradication programme. Yet it was exactly this settlement structure, imposed by the unique coral island environment, that in the end turned out to be beneficial for successful eradication. A distance "of 10 to 20 nautical miles between the islands is rather too long for mosquitoes to cross under normal climatic conditions" and the island communities were sufficiently small to implement efficient systematic blood filming, surveillance and case detection (Schepens 1981, 4).

An earlier road-building programme may also have had some effect by providing useful grounds for facilitating the later WHO eradication programme. In the decade after the Second World War, Muhammad Amin Didi, Prime Minister under the last Sultan and subsequently first president of the Republic of Maldives, ordered the construction of two broad straight roads to cross every island, from east to west and north to south. Therefore, some see Amin Didi as the great modernizer who "aired" the Maldive islands with these avenues and ordered family *veyos*—major breeding grounds for the mosquito species transmitting malaria and lymphatic filariasis (Iyengar et al. 1953, 3; Jambulingam and Krishnamurthy 2014)—to be filled in order to moderate the mosquito burden.[10]

Parasitological and entomological descriptions between the 1950s and 1980s of the small Maldive island habitats as mixed infection zones where humans had to struggle with three different malaria parasites reinforce the historic eye-witness reports of inescapable and deadly fever attacks. With successful eradication, however, the collective memory of Maldive fever faded. Deadly Maldive fever became a scourge of a distant past, banned to the pages of historic reports, to the degree that even the most meticulous of scholars who started fieldwork around the time or after the last indigenous case of severe *P. falciparum* in 1975 considered the Maldive islands to be a place affected merely by benign variants of malaria.[11] As noted above, despite the Maldive fever legacy, the Maldives were largely absent from twentieth-century discourses about malariology and tropical medicine—probably due to comparatively small case numbers. For instance, the malaria distribution maps compiled largely by WHO regional offices (Hay et al. 2004, 328) were based on a minimum threshold of countries and territories with populations above 100,000. Working with precisely

[10] Others, however, link Amin Didi's road construction programme to his "militaristic inclinations" and his desire "to have an avenue in every island to stage parades" with his modernized army (Romero-Frias 2003, 20–23; cf. Maloney 1980, 200–201).

[11] These scholars probably base their opinion on Philip Crowe, the Ambassador of the United States of America to Ceylon. Crowe (1956, 311) reported—at a time when Amin Didi's road constructions might have gained some impact on the islands' mosquito burden—about "some malaria" on the islands, that "only became really dangerous when the health of the people became undermined by starvation". Maloney (1980, 399), however, already briefly reflects anticipatory on the probability of "some degree of genetic adaptation of Divehis [also spelled Dhivehin] to malaria".

these data, the malariologists and cartographers A.Y. Lysenko and I.N. Semashko compiled their seminal geography of malaria, published in 1968. At that time the Maldivian population still was below the threshold figure and thus this heavily burdened archipelago was not included in their study. The more recent "Malaria Success and Challenges in Asia" survey by the WHO South East Asia Regional Office (Bhatia et al. 2013) also ignored the Maldives, thereby missing 600 years of deadly fever history as much as the eradication success in this archipelago.

For the Dhivehin, the dwindling collective memory of massive Maldive fever attacks after malaria eradication was followed by a public health discourse focussing on genetic risk, but which could not offer a causative explanation for the archipelago's contemporary thalassaemia burden. Thus the Dhivehin have come to live in a thalassaemia risk-alert social world that is lacking historic depth. Thalassaemia counselling narratives are disentangled from the causative interactions between environment, culture, biology and evolutionary genetic adaptation. Maldive islanders are struggling with a genetic disposition, yet ignorant of its history. This chapter is an attempt to reassemble the fragmented narratives of Maldive fever, malaria adaptation and eradication, and the fever's aftermath of thalassaemia burdens as a dialectic of *actio* and *passio* (Schnepel 2009), an inextricable entanglement of diverse human and non-human agencies (cf. Tsing forthcoming) in the history of disease dispersion.

Acknowledgements This article is dedicated to the people of the Maldives. It is an attempt to throw light on a distant yet causative past of the health burden with which they live. I am grateful to them for sharing their everyday health problems with me, and thus arousing my academic curiosity. My thanks also extend to Maldive Blood Services (MBS), the Maldivian Thalassaemia Society (MTS) and the Society for Health Education (SHE). I am further grateful to Gwyn Campbell for hosting me as a guest researcher at the Indian Ocean World Centre, McGill University, Christopher Lyons of the Osler Library of the History of Medicine, and Eamon Duffy, University Library, also at McGill, for their guidance. I thank the WHO Archives in Geneva and Reynald Erard for access guidance. I am grateful to Ellen Kattner (Heidelberg) for sharing insights on malaria, mosquito and fever on Maliku and to Mehmet Emir for his kind image processing services. The permission to reprint Picture 11.3 was granted by courtesy of Dr Bernolf Eibl-Eibesfeldt; the image of a Maldivian grass mat (Picture 11.6) was generously provided by KHM-Museumsverband, Welt Museum Vienna. Finally, I also thank the organizers and participants of the conference, as well as Andre Gingrich, Michael Angastiniotis and Boris Wille for valuable remarks on this chapter.

References

Aboobakuru, Maimoona. 2014. Transport Services in the Maldives – An Unmet Need for Health Service Delivery. *Transport and Communications Bulletin for Asia and the Pacific* 84: 15–27.

Adam, Abdulla. 2016. Why Should You Care About Kulhudhuffushi? Climate Change in My Hometown. *SIDS Times* 1(2). http://www.sids2014.org/index.php?page=view&type=12&nr=282&menu=1601. Accessed 1 Sept. 2016.

Ahmad, Waqar I.U., Karl Atkin, and Chamba Rampaul. 2000. 'Causing Havoc Among Their Children?' Parental and Professional Perspectives on Consanguinity and Childhood Disability. In *Ethnicity, Disability and Chronic Illness*, ed. I.U. Ahmad Waqar, 28–44. Buckingham: Open University Press.

Arnold, David. 1991. The Indian Ocean as a Disease Zone, 1500–1950. *South Asia: Journal of South Asian Studies* 14 (2): 1–21. https://doi.org/10.1080/00856409108723152.

Bashyam, Murali D., et al. 2004. Molecular Genetic Analyses of β-thalassemia in South India Reveals Rare Mutations in the β-globin Gene. *Journal of Human Genetic* 49: 408–413. https://doi.org/10.1007/s10038-004-0169-9.

Bell, H.C.P. 1882. *The Máldive Islands: An Account of the Physical Features, Climate, History, Inhabitants, Productions, and Trade*. Colombo: Frank Luker.

———. 1921. *The Màldive Islands: Report on A Visit to Málé, January 20 to February 21, 1920. Ceylon Sessional Papers XV. – 1921*. Colombo: H. R. Cottle, Government Printer, Ceylon.

———. 2002 [1940]. *The Máldive Islands: Monograph on the History, Archaeology and Epigraphy*. Male': Novelty Printers and Publishers.

Bhatia, Rajesh, Rakesh Mani Rastogi, and Leonard Ortega. 2013. Malaria Successes and Challenges in Asia. *Journal of Vector Borne Diseases* 50: 239–247.

Brown, Peter J. 2016 [1981]. Adaptations to Endemic Malaria in Sardinia. In *Understanding and Applying Medical Anthropology*, ed. Peter J. Brown and Svea Closser, 3rd ed., 73–84. London/New York: Routledge.

Cao-Lormeau, Van-Mai. 2016. Tropical Islands as New Hubs for Emerging Arboviruses. *Emerging Infectious Diseases* 22 (5): 913–915. https://doi.org/10.3201/eid2205.150547.

Capua, Aniceta. 1998. The Montreal Thalassemia Screening Program: Response of the High School Students. *Annals of the New York Academy of Sciences* 30: 401–403.

Chatjouli, Aglaia. 2013. Thalassaemic Lives as Stories of Becoming: Mediated Biologies and Genetic (Un)certainties. In *Biosocial Becomings: Integrating Social and Biological Anthropology*, ed. Tim Ingold and Gísli Pálsson, 84–105. Cambridge: Cambridge University Press.

Chiou, Howard. 2016 [2015]. Fundamentals of Evolutionary Medicine. In *Understanding and Applying Medical Anthropology*, ed. Peter Brown and Svea Closser, 3rd ed., 39–46. London/New York: Routledge.

Cooley, Thomas B., and Pearl Lee. 1927. A Series of Cases of Splenomegaly in Children and Peculiar Changes in Bones; Report of Cases. *American Journal of Diseases of Children* 34: 347–363.

Crowe, Philip K. 1956. The Maldive Islands (Chapter xvi). In *Diversions of a Diplomat in Ceylon*, ed. Philip K. Crowe, 280–304. Princeton, New Jersey, New York: D. Van Nostrand Company Inc.. https://ia902701.us.archive.org/27/items/diversionsofadip007986mbp/diversionsofadip007986mbp.pdf. Accessed 26 July 2016.

Dunn, Ross E. 2005 [1986]. *The Adventures of Ibn Battuta. A Muslim Traveler of the 14th Century*. Berkeley: University of California Press.

Dyson, Simon M., I.U. Waqar Ahmad, and K. Atkin. 2016. Narrative as Re-fusion: Making Sense and Value from Sickle Cell and Thalassaemia Trait. *Health: An Interdisciplinary Journal for the Social Study of Health, Illness and Medicine* 20 (6): 616–634.

Firdous, Naila. 2005. Prevention of Thalassaemia and Haemoglobinopathies in Remote and Isolated Communities: The Maldives Experience. *Annals of Human Biology* 32 (2): 131–137.

Firdous, Naila, Stephen Gibbons, and Bernadette Modell. 2011. Falling Prevalence of Beta-Thalassaemia and Eradication of Malaria in the Maldives. *Journal of Community Genetics* 2: 173–189.

Fitzler, M. A. Hedwig. 1936. Die Malediven im 16. und 17. Jahrhundert: Ein Kapitel portugiesischer Kolonialgeschichte. *Zeitschrift für Indologie und Iranistik* 10: 215–256.

Forbes, Andrew D.W. 1981. Southern Arabia and the Islamicisation of the Central Indian Ocean Archipelagoes. *Archipel* 21: 55–92.

———. 1987. The Pre-Islamic Archaeology of the Maldive Islands. *Bulletin de l'Ecole fançaise d'Extrême-Orient* 76: 281–288.

Forbes, Andrew, and Fawzia Ali. 1980a. *Weaving in the Maldive Islands, Indian Ocean: The Fine Mat Industry of Suvadiva Atoll: Occasional Paper No 9*. London: British Museum.

———. 1980b. The Maldive Islands, Indian Ocean, and their Historical Links with the Coast of Eastern Africa. *Kenya Past and Present* XII: 15–20.

Furuumi, H., et al. 1998. Molecular Basis of β-Thalassemia in the Maldives. *Hemoglobin* 22 (2): 141–151.

Geertz, Clifford. 1983. *Local Knowledge: Further Essays in Interpretive Anthropology*. New York: Basic Books.

Greene, Jeremy, Marguerite T. Basilico, Heidi Kim, and Paul Farmer. 2013. Colonial Medicine and Its Legacies. In *Reimagining Global Health: An Introduction*, ed. Paul Farmer, Jim Yong Kim, Arthur Kleinman, and Matthew Basilico, 33–74. Berkeley: University of California Press.

Haldane, J.B.S. 1949. The Rate of Mutation of Human Genes. *Hereditas* 35 (S1): 267–273.

Hay, Simon I., et al. 2004. The Global Distribution and Population at Risk of Malaria: Past, Present, and Future. *The Lancet Infectious Diseases* 4 (6): 327–336. https://doi.org/10.1016/S1473-3099(04)01043-6.

Hedrick, P.W. 2011. Population Genetics of Malaria Resistance in Humans. *Heredity* 107: 283–304.

Hewitt, Rachel, et al. 1996. β-Globin Haplotype Analysis Suggests That a Major Source of Malagasy Ancestry is Derived from Bantu-Speaking Negroids. *American Journal of Human Genetics* 58: 1303–1308.

Hockly, T.W. 1935. *The Two Thousand Isles: A Short Account of the People, History and Customs of the Maldive Archipelago*. London: H.F. & G. Witherby.

Hogendorn, Jan, and Marion Johnson. 1986. *The Shell Money of the Slave Trade*. Cambridge: Cambridge University Press.

Ibrahim, Shaheedha Adam, Mohamed Rasheed Bari, and Leon Miles. n.d. *Water Resources in Maldives with an Emphasis on Desalination*. Male': Maldives Water and Sanitation Authority. http://citeseerx.ist.psu.edu/viewdoc/download?doi=10.1.1.113.913&rep=rep1&type=pdf. Accessed 30 May 2013.

Iyengar, M.O.T. 1952. Filariasis in the Maldive Islands. *Bulletin of the World Health Organisation* 7: 375–403.

Iyengar, M.O.T., and M.A.U. Menon. 1955. Mosquitos of the Maldive Islands. *Bulletin of Entomological Research* 46: 1–10.

Iyengar, M.O.T., M.I. Mathew, and M.A.U. Menon. 1953. Malaria in the Maldive Islands. *Indian Journal of Malariology* 7 (1): 1–3.

Jambulingam, P., and K. Krishnamurthy. 2014. Elimination of Lymphatic Filariasis in Maldives: A Success Story. *Health in South-East Asia. SEARO Newsletter* 7 (1): 12–13.

Johnson, Donald R. 1966. Progress Toward Malaria Eradication in Asia. *Mosquito News* 26 (4): 476–482.

Kattner, Ellen. 2007. Bodu Valu – Big Ponds. Traditional Water Management and its Socio-cosmic Implications in Minicoy/Maliku, an Indian Ocean Island. In *Antike Zisternen*, ed. Christoph Ohlig, 145–172. Siegburg: Deutsche Wasserhistorische Gesellschaft (DWhG) Bd. 9.

Kläy, Ernst Johannes. 1986. Die Republik der Malediven. In *Trauminseln – Inselträume. Die Republik der Malediven (Indischer Ozean) im Spiegel westlicher Vorstellungen*, ed. Ernst Johannes Kläy and Daniel Kessler, 59–154. Bernisches Historisches Museum: Bern.

Knoll, Eva-Maria. 2018a. The Maldives as an Indian Ocean Crossroads. In *Oxford Research Encyclopedia of Asian History*. New York: Oxford University Press. Article published October 2018. https://doi.org/10.1093/acrefore/9780190277727.013.327.

———. 2018b. Considering the Island Capital Male' as a Hub for Health-related Mobilities. In *Connectivity in Motion: Island Hubs in the Indian Ocean World*, ed. Burkhard Schnepel and Edward Alpers, 389–419. Basingstoke and New York: Palgrave Macmillan.

Lévi-Strauss, Claude. 1966. *The Savage Mind*. University of Chicago Press.
Maloney, Clarence. 1980. *People of the Maldive Islands*. Orient Longman.
Maniku, Hassan Ahmed. 1988. *Dhevi*. Male': Department of Information and Broadcasting.
Mitchell, John J., et al. 1996. Twenty-Year Outcome Analysis of Genetic Screening Programs for Tay-Sachs and β-Thalassemia Disease Carriers in High Schools. *American Journal of Human Genetics* 59: 793–798.
MoH – Ministry of Health. 2015. National Health Research Priority Areas. *Maldives Health Research Bulletin* 2015 (2): 20. Male'.
Mohamed, Naseema. 2008. *Essays on Early Maldives*. Male': National Centre for Linguistic and Historical Research.
O'Brien, John. 2011. Bats of the Western Indian Ocean Islands. *Animals* 1 (3): 259–290. https://doi.org/10.3390/ani1030259.
Ottovar, Annagrethe, and Nils Finn Munch-Petersen. 1980. *Maldiverne—et øsamfund i det Indiske Ocean*. Copenhagen: Kunstindustrimuseet.
Packard, Randall M. 2007. *The Making of a Tropical Disease: A Short History of Malaria*. Baltimore: The Johns Hopkins University Press.
Pearson, Michael. 2003. *The Indian Ocean*. London and New York: Routledge.
Piel, Frédéric B., et al. 2010. Global Distribution of the Sickle Cell Gene and Geographical Confirmation of the Malaria Hypothesis. *Nature Communications* 1: 104. https://doi.org/10.1038/ncomms1104.
Pijpe, Jeroen, et al. 2013. Indian Ocean Crossroads: Human Genetic Origin and Population Structure in the Maldives. *American Journal of Physical Anthropology* 151: 58–67.
Prainsack, Barbara, and Gil Siegal. 2006. The Rise of Genetic Couplehood? A Comparative View of Premarital Genetic Testing. *BioSocieties* 1: 17–36.
Pratt, Mary Louise. 1986. Fieldwork in Common Places. In *Writing Culture. The Poetics and Politics of Ethnography*, ed. James Clifford George Marcus, 27–49. Berkeley, Los Angeles, London: University of California Press.
Ptak, Roderich. 1987. The Maldive and Laccadive Islands (liu-shan 溜 山) in Ming Records. *Journal of the American Oriental Society* 107 (4): 675–694.
Pyrard, François. 1887. *The Voyage of François Pyrard of Laval to the East Indies, the Maldives, the Moluccas, and Brazil*. Translated and edited with notes by Albert Gray, assisted by H.C.P. Bell from the 3rd French edition of 1619, Vol. I. New York: Burt Franklin Publisher.
Raben, Remco. 1996. European Periphery at the Heart of the Ocean: The Maldives, 17th–18th Centuries. In *International Conference on Shipping, Factories and Colonization (Brussels, 24–26 November 1994)*, ed. John Everaert and J. Parmentier, 45–60. Collectanea maritima Bd. 7. Brussel: Académie Royale des Sciences d'Outre-Mer.
Romero-Frias, Xavier. 2003 [1999]. *The Maldive Islanders: A Study of the Popular Culture of an Ancient Ocean Kingdom*. Barcelona: Nova Ethnographia Indica.

———. 2012. *Folk Tales of the Maldives*. Copenhagen: NIAS.

Rosset, Carl W. 1886/1887. Carl Rosset in Male', Maldives 1885 – Original Photos and Lithographs. Reprint from *The Graphic* 16 October 1886, with additions from *Illustrierte Zeitung* 23 April 1887. Trans. Lars Vilgon. http://www.maldivesculture.com. Accessed 4 Oct 2014.

Rosset, C.W. 1887. On the Maldive Islands, More Especially Treating of Male Atol. *The Journal of the Anthropological Institute of Great Britain and Ireland* 16: 164–174.

Royal Naval Medical Services. 1954. *Volume I, Administration*. By Surgeon Commander J. L. S. Coulter, D.S.C., R.N., Barrister-at-Law. London: Her Majesty's Stationery Office.

Schepens J.R.M. 1981. *Malaria Control in the Maldives: WHO Consultant's Assignment Report, October 1974–July 1980*. WHO Project MAV MPD 001.

Schnepel, Burkhard. 2009. Zur Dialektik von *agency* and *patiency*. In *Handlung und Leidenschaft. Jenseits von actio and passio*, ed. Klaus-Peter Köpping, Burkhard Schnepel, and Christoph Wulf. *Paragrana. Internationale Zeitschrift für Historische Anthropologie* 18(2): 15–22.

Scriver, Charles R., et al. 1984. β-Thalassemia Disease Prevention: Genetic Medicine Applied. *American Journal of Human Genetics* 36: 1024–1038.

Tsing, Anna L. forthcoming. Organisms as Agility-Shifting Actors in the Trajectory of Capital: The 11th Eric Wolf Lecture. *Current Anthropology*.

Velimirovic, Boris, and J.L. Clarke. 1975. Use of Larvivorous Fish for Vector Control in the Maldives Republic. *Tropical Medicine and Parasitology* 26: 503–506.

Victor, Isaac H. 2014. Maldives. Revised by Xavier Romero-Frias. In *Worldmark Encyclopedia of Religious Practices*, 2nd ed., Vol. 3, 378–386. Gale.

Viganó, Claudia, et al. 2017. A 2,000 Year Old Β-thalassemia Case in Sardinia Suggests Malaria was Endemic by the Roman Period. *American Journal of Physical Anthropology* 164: 362–370. https://doi.org/10.1002/ajpa.23278.

Vilgon, Lars. 2001a. 1344. Ibn Battuta. In *Maldive Odd History, Volume I: A Collection of 44 Entries Translated and Transliterated into English from 11 Languages*, ed. Lars Vilgon, 25–53. Male': National Centre for Linguistic and Historical Research.

———. 2001b. 1835: Campbell, David. In *Maldive Odd History, Volume IV: A Collection of 13 Entries Translated and Transliterated into English from 7 Languages*, ed. Lars Vilgon, 63–103. Male': National Centre for Linguistic and Historical Research.

Villiers, John. 1992. The Portuguese in the Maldive Islands. In *Studies in the Portuguese Discoveries I. Proceedings of the First Colloquium of the Centre for the Study of the Portuguese Discoveries*, ed. T.F. Earle and Stephen Parkinson, 17–34. Warminster: Aris & Phillips.

Weatherall, David. 1997. Thalassaemia and Malaria, Revisited. *Annals of Tropical Medicine and Parasitology* 91 (7): 885–890.
———. 2010. *Thalassaemia. The Biography*. Oxford: Oxford University Press.
Webb, Paul A. 1988. Economy of the Islands. In *Maldives. People and Environment*, ed. Paul A. Webb, 33–39. Bangkok: Media Transasia Limited.
Webb, James L.A., Jr. 2009. *Humanity's Burden: A Global History of Malaria*. Cambridge: Cambridge University Press.
WHO. 2016. *Malaria-Free Maldives*. New Delhi: WHO Regional Office for South-East Asia.
Yoosuf, Abdul Sattar. 2014. Eliminating Malaria in Maldives. *Health in South-East Asia. SEARO Newsletter* 7 (1): 8–10.

Index[1]

[1] Note: Page numbers followed by 'n' refer to notes.

G. Campbell, E.-M. Knoll (eds.), *Disease Dispersion and Impact in the Indian Ocean World*, Palgrave Series in Indian Ocean World Studies, https://doi.org/10.1007/978-3-030-36264-5

D

E

I

J

The manufacturer's authorised representative in the EU is Springer Nature Customer Service Centre GmbH, Europaplatz 3, 69115 Heidelberg, Germany. If you have any concerns regarding our products, please contact ProductSafety@springernature.com

Printed and bound by CPI Group (UK) Ltd, Croydon, CR0 4YY
30/09/2026
02225644-0010